Joseph Prestwich

Collected Papers on Some Controverted Questions of Geology

Joseph Prestwich

Collected Papers on Some Controverted Questions of Geology

ISBN/EAN: 9783337218553

Printed in Europe, USA, Canada, Australia, Japan

Cover: Foto ©berggeist007 / pixelio.de

More available books at **www.hansebooks.com**

COLLECTED PAPERS

ON SOME

CONTROVERTED QUESTIONS OF GEOLOGY

COLLECTED PAPERS

ON SOME

CONTROVERTED QUESTIONS

OF GEOLOGY

BY

JOSEPH PRESTWICH, D.C.L. (Oxon.), F.R.S., F.G.S.,

CORR. INST. FRANCE (ACAD. SCI.); ACAD. R. LYNCEI, ROME; IMP. GEOL. INST. VIENNA
ACAD. ROY. BRUSSELS; AMER. PHIL. SOC. PHILAD.; ETC., ETC.

London
MACMILLAN AND CO.
AND NEW YORK
1895

The Right of Translation and Reproduction is Reserved

RICHARD CLAY AND SONS, LIMITED,
LONDON AND BUNGAY

PREFACE

WITH respect to the main facts of Geology, we geologists are in general of one opinion, but with respect to the explanation of many of those facts, we hold very divergent opinions. These opinions, which go to the very foundation of theoretical geology, depend mainly upon the school to which the writers belong.

In England, the prevailing school for many years past is the one which teaches the doctrine of Uniformity—Uniformity of action, both in *Kind and in Degree* throughout all geological time. The other school which has fewer adherents in this country, but is the larger one on the Continent, inculcates Uniformity in *kind or law*, but not *Uniformity in degree*. These questions are briefly reviewed in the Article No. 1.

The application of *Uniformity of degree*, in conjunction with Croll's chronological measures, has led to the assigning to the Glacial Period and to the Antiquity of Palæolithic Man a length of time greater than the geological evidence would seem to warrant. This question is discussed in the second paper.

On the other hand, there has been an unwillingness to apply the same rules to the Antiquity of *Plateau Man*. The latter, according to some Geologists, would be no older than Palæolithic Man of the *Valley gravels*. The reasons for a greater Antiquity and for the genuineness of the Plateau implements are given in the third Article.

The presence of water in Volcanic Eruptions is another subject on which a difference of opinion as to its origin and effects, exists. The cause of the vast escape of steam from Volcanoes has been held to be due to steam held or occluded, wholly or in great part, in the lava in the Volcanic foci. The object of the fourth Article is to show on the contrary that lava in the Volcanic foci does not contain occluded water, but that the waters which give rise, as well to the escape of steam as to explosions, are surface waters, which in their descent come in contact with the lava in the duct of the volcano as it approaches the surface.

With respect to the thickness of the Earth's Crust, very different opinions are held by Physicists and Geologists. The reasons on either side are stated briefly in the fifth Article. The geological evidence is in favour of a thin crust, but of a solid nucleus.

The adoption of a mean thermometrical gradient, or rate of increase in feet for each degree of temperature, was long and possibly may yet be considered an unsettled problem. A rate of 50 feet for a degree has been sometimes adopted, though not on any fixed principle, while by some geologists rates of 32 to 70, of 37 to 41, of 46 to 51, and of 60 to 64, have been adopted. This difference may be partly, but is not wholly accounted for by the variation in the geological structure and by the nature of the work. Former lists give only from 30 to 60 observations. The list here appended comprises 231 observations, which were scattered through many periodicals. Some of the added observations materially affect the conclusions respecting the thermometric gradient. This forms the subject of the sixth Article.

In Geology, as in other subjects, there are commonly two sides to most questions. One of these has of late years mainly monopolised attention, to the neglect of the other. In these papers, I have given prominence to the latter, in the hope that it may lead to a fresh discussion of the subject.

I have to express my obligation to the Editor of the *Nineteenth Century* for permission to reprint the first of these papers, and to the Presidents and Councils of the Royal, Geological, and Anthropological Societies for the same permission with respect to the others.

The figures of the Plateau Implements have been rendered with much fidelity and skill by Mr. W. S. Tomkin.

CONTENTS

I. ARTICLE

PAGE

THE POSITION OF GEOLOGY (a Chapter on Uniformitarianism). *Nineteenth Century* for October, 1893, p. 551 1

II. ARTICLE

CONSIDERATIONS ON THE DATE, DURATION, AND CONDITIONS OF THE GLACIAL PERIOD, WITH REFERENCE TO THE ANTIQUITY OF MAN (REVISED). *Quart. Journ. Geological Society* for August, 1887, vol. 43, p. 393 19

   1. Extreme Estimates of Time—2. Croll's Hypothesis considered—3. Rates of Glacier Motion—4. The Alps; Greenland—5. Climatal Effects of Submergence—6. Excavation of Valleys—7. High- and Low-Level, and Glacial Gravels—8. Date of Palæolithic Man.

III. ARTICLE

ON THE PRIMITIVE CHARACTERS OF THE FLINT IMPLEMENTS OF THE CHALK PLATEAU OF KENT, WITH REFERENCE TO THE QUESTION OF AGE AND MAKE (REVISED WITH ADDITIONS). *Journ. Anthropological Institute* for 1892, p. 246 49

   1. Position of the Plateau Implements—2. The Geological Question—3. Alternative Explanation of their Origin—4. Distinctive Characters of the Valley and Plateau Implements. The Valley Implements. The Plateau Implements—5. Typical Forms of the Plateau Implements.

IV. ARTICLE

ON THE AGENCY OF WATER IN VOLCANIC ERUPTIONS, AND ON THE PRIMARY CAUSE OF VOLCANIC ACTION (REVISED). *Proceedings of the Royal Society*, April, 1885, p. 117 81

   1. Introductory Observations—The Occluded Vapour of Water considered as the Primary Cause of Volcanic Eruptions—

2. Objections to this Hypothesis—3. Influence of Volcanic Eruptions on Spring and Well Waters—4. The Hydro-geological and Statical Condition of the Underground Waters in and under a Volcanic Mountain in a State of Rest—5. Condition of the Underground Waters during an Eruption—6. Transmission of the Surface Waters into the Volcanic Duct—7. The Ejection of Blocks of Rock during Eruptions—8. Primary Cause of Volcanic Action.

V. ARTICLE

ON THE THICKNESS AND MOBILITY OF THE EARTH'S CRUST FROM THE GEOLOGICAL STANDPOINT (REVISED). *Proceedings of the Royal Society*, April, 1888, p. 156 147

1. Flexibility of the Crust—2. Extent of Mountain Compression—3. Effects of a Thick Crust—4. Measured by the Broken Edges—5. Probable Thickness—6. The Effects of Conductivity on Temperature—7. Volcanic Action incompatible with a Thick Crust.

VI. ARTICLE

ON UNDERGROUND TEMPERATURES; WITH OBSERVATIONS ON CERTAIN CAUSES WHICH INFLUENCE THE CONDUCTIVITY OF ROCKS; ON THE THERMAL EFFECTS OF SATURATION AND IMBIBITION; AND ON A SOURCE OF HEAT IN MOUNTAIN RANGES, AS AFFECTING SOME UNDERGROUND TEMPERATURES (REVISED). *Proceedings of the Royal Society*, February, 1885, p. 1 166

1. Early Methods—2. Interfering Causes—3. General List of Observations in Order of Date—4. Observations in Coal Mines—5. Observations in Mines other than Coal—6. Artesian Wells and Borings—7. Tunnels—8. Conductivity of Rocks—9. Effects of Saturation and Imbibition—10. Conclusions.

LIST OF ILLUSTRATIONS

III. ARTICLE

                                                      PAGE

FIG. 1.—DIAGRAM SHOWING THE RELATIVE POSITIONS OF THE IMPLEMENT-BEARING DRIFT ON THE CHALK PLATEAUS AND OF THE VALLEY DRIFTS OF POSTGLACIAL AGE .   53

PLATES I. to XII.—TYPES OF THE PLATEAU IMPLEMENTS
                                              To face page   80

IV. ARTICLE

FIG. 1.—DIAGRAM SECTION OF THE WATER-LEVEL UNDER ETNA   124

FIG. 2.—SECTION ON THE SLOPES OF THE OLD VOLCANO OF SANTORIN (FOUQUÉ)   128

PLATE XIII.—DIAGRAM SECTION OF A VOLCANO IN ERUPTION
                                              To face page   130

VI. ARTICLE

FIG. 1.—SECTION OF UPPER DUFFRYN COLLIERY, ABERDARE DISTRICT   196

FIG. 2.—SECTION OF CYM NEOL COLLIERY, ABERDARE DISTRICT   196

FIG. 3.—SECTION OF NEW TREDEGAR COLLIERY, ABERDARE DISTRICT   197

FIG. 4.—SECTION OF VOCHRIW DOWLAIS COLLIERY, ABERDARE DISTRICT   197

FIG 5.—SECTION ACROSS ST. GOTHARD (reduced from the large Section by Dr. Stapff)   236

I

THE POSITION OF GEOLOGY

THE position of Geology in this country at the present time, more especially as relates to the later geological periods, is anomalous and possibly without precedent. On one side its advance is barred by the doctrine of Uniformity, and on the other side by the teaching of Physicists. The former requires that everything should be regulated by a martinet measure of time and change. It asserts that the vast changes on the earth's surface, effected during long geological periods, are to be *measured by the rate at which similar but minor changes are effected in the present day*, and that the agencies now modifying the surface have been alike, in every respect, in all past time.[1] It is true that no restriction is placed on the extent of the changes, but such prolonged time is insisted on for their accomplishment as to destroy the value of

---

[1] We cannot admit the exception claimed for Volcanic action. One form of it was certainly more active in the past than at present, while on the other hand, it is possible that "explosive eruptions" are more violent now than in former times.

the concession. Not that time is in itself a difficulty, but a time-rate, assumed on very insufficient grounds, is used as a master-key, whether or not it fits, to unravel all difficulties. What if it were suggested that the brick-built Pyramid of Hawâra had been laid brick by brick by a single workman? Given time, this would not be beyond the bounds of possibility. But Nature, like the Pharaohs, had greater forces at her command to do the work better and more expeditiously than is admitted by Uniformitarians.

On the other side, Physicists would lead us to suppose that those great movements of the earth's crust, with which we are all familiar in the form of high mountain and continental upheavals in the earlier stages of the Earth's history, were impossible in those times which more immediately approached our own. They formerly maintained that its outer crust at least had a thickness estimated at from 800 to 2,500 miles. Now, however, they regard it as solid throughout, and so rigid that we are forced to believe that for a long preceding period it must have been in a state of comparatively stable equilibrium. This, however, would have rendered the great earth movements, considered by Geologists to have continued up to the threshold of our own times, impossible. And to this finding the Physicists would have geological speculations conform. At the same time, judging, amongst other reasons, from the rate of cooling of hot solid bodies, they would assign a much shorter term to the earth's history *since it became habitable* than is compatible with the views of the Uniformitarian school of Geologists. The one side counts in round numbers upon some three hundred million years; the other sees no reason to go beyond fifteen to twenty million years—a term, in our humble opinion, much more probable than the other.

On another point, our two allies (allies in the sense of working at the same subject) are in irreconcilable antagonism. The Physicists tell us that uniformity of action in all time is impossible, while the Uniformitarians say that such a shortening of geological time as would follow on the acceptance of the physical argument is against all geological experience. Not only do these opinions clash, but those also concerning the rigidity of the earth and the thickness of its crust are widely divergent. None of these contentions can, however, be disregarded, for we must all recognise the importance of considering the question from every point of view. The argument in favour of uniformity of action has been put before us with so much skill and ability, and possessing as it does the charm of an infallible faith, that Uniformitarianism has become the accepted doctrine of the dominant school of geology. Besides, within certain limits and in certain lights, the arguments of the Uniformitarian and of the Physicist might hold good—that is to say, if we would restrict the deductions of the former to the recent period, and could adopt the propositions of the latter. Our part, however, is to see whether their conclusions agree —not with their respective assumptions, but with the geological evidence: for no conclusions can be accepted that do not meet with the full concurrence of all the co-partners interested in the result, and without respect for their mutual claims progress is not possible. The Geologist must attend to the claims of the Physicist, and the Physicist should not overlook those of the Geologist. How then stands the case?

With regard to the geological problem, we are told by the Uniformitarians that the forces acting on the surface of the globe have been in all past times the same, both in *kind* and *degree,* as those now in operation. On those grounds they would estimate, first, the time required for mountain and continental elevation; secondly, the rate of erosion of the valleys, and, thirdly, of the denudation or lowering of the land. Their conjecture is that our limited experience of 2,000 to 3,000 years has sufficed to furnish us with instances of all the various vicissitudes and changes that the earth has undergone during the illimitable past—a generalisation incompatible with what is known of the evolution of the earth, and in contradiction to their own premisses. For even Geologists who recognise no change admit the original molten state of the globe. This of itself involves, in the cooling of the mass, the intervention of stresses and strains, with all their consequences, which render it inconceivable that there was nothing in all those stages of the earth's history beyond what our limited experience has brought us in contact with.

But although the assumption of the Uniformitarians on the question of *degree* may be disputed, that on the question of *kind* admits of no dispute. That rivers excavate and currents distribute the excavated materials, and that the land is mobile and subject to changes of level, no one will contest. The point of contention is the *rate* at which these operations and changes proceeded formerly as compared with the rate at the present day. The many observations made on the erosive and transporting power of rivers, and on the movements and waste of the land, are admirable in so far as they apply to the silting up of ports, the recession of the coast, and the reclamation of marsh lands; but, though valuable to the engineer, they are misleading to the Geologist. They furnish him, it is true, with standards applicable to *present* changes, and indicate *the method* in which the erosive power of the rivers and seas has acted in all time, but they give no measure of the *amount and rate of work* they did at different periods. Nevertheless, knowing what at present is accomplished by their means, it is reasonable to judge, by ascertaining what their agency accomplished in former days, of the difference in the activity of the forces in operation at the several periods. Those forces have to be estimated by the work done in the past, and not by any fixed rate founded upon present work.

Few Geologists would, we presume, contest this position; notwithstanding which, and though some now profess a modified Uniformitarianism, the old lines of argument still, with few exceptions, prevail, and the concessions made are more apparent than real, or are of little value. In our opinion, no partial concession can be entertained on the question of *degree*. It must be an unconditional surrender; for, in contradistinction to *kind or law*, where we are on common ground, no common scale on the question of *degree* is possible in judging of the past by comparison with the present.[1]

As an example of the present position, we may take one argument as presented by the advocates of the Uniformitarian school. The observations on the transporting power of the large rivers of the world have shown that the quantity of sediment carried down by them to the sea is, according to an estimate made by them, such as would suffice to lower the level of the land about one foot in 6,000 years, or about 1,000 feet in 6,000,000 years. Exception might be taken to this estimate in that no account is taken of the calcareous matter removed in solution, which, in fact, is not far from the quantity of insoluble matter carried down mechanically. Let that pass. This measure, or one approximate to it, has been generally accepted, and is in common use. Hence, Geologists of that school, proceeding solely on the assumed postulate, and not attaching due

---

[1] A little more than a quarter of a century ago hardly a voice was raised (with the exception of that of Sir R. Murchison, Prof. Sedgwick, and a very few others besides myself) against the doctrine of Uniformity in all time, which still remains the creed of the majority, though I believe, in many cases, this arises from confounding *degree* with *kind*. The explicit statement of the Director-General of the Geological Survey, Sir Archibald Geikie, in his Presidential Address at the meeting of the British Association in Edinburgh in 1892, affords satisfactory evidence of the turn in the tide.

weight to other considerations, have, it seems to us, placed the later Quaternary times at far too great a distance from the present. In the same way, the rate at which the elevation of the land took place having been estimated on the mean of $2\frac{1}{2}$ feet in a century, would, if that scale were accepted, manifestly push back to a very remote distance even later geological changes of level.

The importance of determining these points more accurately became more evident when it was discovered that Man existed with the extinct Mammalia; and therefore upon the solution of the time-rate problem depended the determination of the antiquity of Man upon the earth. Various have been the attempts since made; but, as they have almost all been made upon measurements based on the above-named scales, they necessarily involved a very free use of time. For long, Geologists had held to the belief, prevailing half a century ago, that Man could not have existed on the earth for more than 5,000 to 6,000 years. When evidence was given, and at last accepted, to prove a higher antiquity, the Uniformitarians were placed in the difficulty of proving too little or too much. If they adopted a short chronology, it would clash with the corner-stone of their belief as to the age of the Quaternary deposits; if, on the other hand, they retained their belief in the great length of time they held to be necessary for the formation of the post- or later-glacial deposits, they would have to assign to Man an antiquity which would clash sorely not only with their own previous belief, but also with that held on various grounds by some of their own school and by Anthropologists.

The *Fetish* of uniformity prevailed, the Uniformitarians made *volte-face* to their former contention, and hesitated not, at first, to claim for Man an antiquity of nearly a million years. One friend of ours in a public lecture, even put in a claim for two millions, heedless of the cries of his unprepared audience to remind him of the rights of Adam. At a loss to prove their case by independent geological evidence, they found an unexpected ally in a novel and ingenious astronomical hypothesis, which apart from its connection with geology we will not contest.[1] The object of the hypothesis was to show that there had been cycles, in which at times the position of the earth in its orbit was such as would cause a great lowering of the terrestrial temperature, and give rise to recurring Glacial periods. Here were offered the definite measures that geology failed to furnish, and which tallied too well with the time needed by the Uniformitarians to be neglected. It was therefore eagerly adopted, and has since been prominent in geological literature. That the hypothesis, however, is not in accordance with the facts of geology has been abundantly shown both in America and in this country; nevertheless the belief prevails. The result is that, as the last of these astronomical periods was calculated to have commenced 250,000 years and to have ended 80,000 years ago, these

---

[1] On other grounds it is possible that the antiquity of Man will have to be carried back further into the Glacial period. See Paper No. III. of this volume.

numbers have become stereotyped as those of the beginning and the end of the Glacial period.[1]

The able author of this hypothesis, in his attempt to reconcile geological and astronomical time, built his geological argument upon the rate of erosion of rivers at the present time, as held by the Uniformitarian.[2] Nevertheless, an observation of his own, that must be endorsed by all Geologists, whatsoever their creed, shows the fallacy of adopting the rates of the present day as measures for the past, for he remarks, "*if the rate of denudation be at present so great, what must it have been during the Glacial period? It must have been something enormous.*" Very true, yet the argument proceeds as before. With the admission here made, how is it possible to adopt a scale admitted by its advocate to be subject to such variation? Its retention only serves to divert the real issue and stay inquiry.[3]

Another objection to this chronology is that it fixes the date of the disappearance of Palæolithic Man and the Quaternary fauna at a distance of 80,000 years from our own times. Of these 80,000 years, we can account for 10,000 or 12,000, during which Neolithic and recent Man has been in occu-

---

[1] See Paper No. II.

[2] *Climate and Time*, chapter xx.

[3] The rate of denudation at the different geological periods must necessarily depend mainly on the rainfall. Of this we know nothing. We can only judge by the results, which indicate it to have been excessive at many periods. It is useless, therefore, to attempt to measure geological time by the rate at which sedimentation now takes place. This may serve as a point of comparison, but nothing more.

pation of the land; but this leaves some 70,000 years unaccounted for. Unable satisfactorily to show on geological grounds the need of so great an interval between the end of the Quaternary period and the present time, the Uniformitarians find a more colourable defence on biological grounds. They point, in a manner we do not quite understand, to the circumstance that with the close of the Post-glacial period a number of the animals then living disappear from the scene, and contend that for the dying out of so many species long ages must have been required. Had they been able to show the working of evolution in the coming in of new species by descent from the extinct species, or of change in the contemporary species still living, their argument could not be gainsaid. But there is no question of evolution. The Mammoth and woolly Rhinoceros disappeared comparatively suddenly and leaving no descendants; the Reindeer, Musk-ox, and Glutton were driven to northern latitudes, and there still survive unchanged; while the Horse, Ox, Red-deer, Wolf, Bear, and others remained on with us without variation. The extraordinary change of climate which then took place is quite sufficient to account for such changes as these, which are chiefly those of faunal distribution, having been effected in a measurable length of time, instead of needing the vastly long period mentioned. For this length of time, even without the aid of the physical changes which then took place, could hardly have failed to involve more extensive changes in the species than are apparent in the surviving species.

There is, in fact, no sufficient evidence, either geological or biological, to show the need of the long interval assumed. On the contrary, there is every reason to believe that it did not exist, but that Palæolithic Man and his companions came down to within some 10,000 to 12,000 years of our times. We cannot suppose that either Man or geological processes would have remained stationary during 70,000 years, and yet that is the conclusion we should be driven to adopt. Are we to be debarred from pursuing these inquiries by an astronomical hypothesis having no better foundation, and involving such unquestionable difficulties?

Another barrier to inquiry is the postulate which would fix the rate of upheaval of the land during geological periods upon observations based—not upon the experience of even 2,000 or 3,000 years—but upon observations which do not extend beyond two centuries. These observations have shown that the *mean* rate of elevations of the coasts of Norway and Sweden has been during that time $2\frac{1}{2}$ feet in a century, and this scale has been accepted and commonly employed by the Uniformitarian School as a safe and sure basis for calculation of geological time. The determination of a secular rise of the land is of itself an interesting fact, as settling the question of a retained mobility in the earth's crust; but it is quite insufficient, even if it were applicable, to establish a definite rate, not only for the past but even for the present. It is not a mean rate that is wanted. No upheaval can be otherwise than local and graduated. The extremes are what are needful. No engineer would take the mean delivery of a river as the measure to be depended upon for a water-supply. It is the limit in both directions, or the minimum and maximum quantities, that are essential. To know what earth movements can still effect, Uniformitarians should at least take the maximum rate, which amounts in the above case, at the North Cape, to 5 feet in the century, or double the measure of the mean adopted by them.

If also, in calculating the present rate of elevation of the land, the mean rate along the whole length of the axis is adopted, the same rule should at least be applied to elevations of past periods, and the time should not be estimated by the height of any one point, as that may prove to be more or less in excess of the mean. Thus, for example, the Westleton marine shingle is found in Buckinghamshire at a height of 600 feet. Estimating this upheaval at the rate of $2\frac{1}{2}$ feet in a century, the Uniformitarian would put in a claim for 24,000 years. But this bed, as it trends eastward, is met with at gradually lower levels, until in Suffolk it falls to the sea-level. A mean of 300 feet should therefore be taken, with a corresponding shorter time-term of 12,000 years; or are we to look only at the beds on the coast, where they are nearly in their original place and level? From every point of view such estimates must be worthless.

More than this, the very leaders of belief that the average rate of motion does not exceed that above named do not deny that "*the average rate proposed is a purely arbitrary and conjectural one.*" It is admitted also that it is not improbable that during the last 400 years there has been a still faster rate in high northern latitudes. Not only, however, is the half-measure adopted, but the warning that higher measures exist is neglected. When therefore the mean is applied to determine the length of time required to effect such elevations as that of the marine shell bed on Moel Tryfaen, 1,400 feet above sea-level and of late Quaternary age, Uniformitarians are obliged to ask for a term of 56,000 if not 88,000 years.[1] Should the case of Moel Tryfaen be objected to as uncertain, there are still the unquestioned Raised Beaches of Norway and Sweden, which are from 200 to 600 feet above the sea-level, and of still more recent date. These, on the same estimate, would have taken for their upheaval some 8,000 to 24,000 years. We need not, however, pursue this subject further. The very admissions made by the advocates of the two above-named measures of time, based upon present rates of denudation and of elevation, show how untenable their conclusions are.

Such observations, howsoever useful and suggestive, are in fact futile so far as regards their application to former rates of upheaval, and needlessly play with time. If we could suppose that the causes which produced those movements had always acted with the same degree of energy the reasoning would hold good; but, as that regularity depends upon the stresses to which the earth's crust has been exposed at any particular time, the effects must have varied

[1] *Antiquity of Man,* 4th edit., p. 335.

in proportion as the stresses varied. With a cooling globe it could not, it seems to us, have been otherwise. What those movements of the past were, and what their duration, must therefore be judged of by other circumstances and on surer data.

We trust we have now said enough to show upon how insecure a basis the Uniformitarian measures of time and change stand. They have probably done as much to impede the exercise of free inquiry and discussion as did the catastrophic theories which formerly prevailed. The latter found their own cure in the more accurate observation of geological phenomena and the progress of the collateral sciences; but the former hedge us in by dogmas which forbid any interpretation of the phenomena other than that of fixed rules which are more worthy of the sixteenth than of the nineteenth century. Instead of weighing the evidence and following up the consequences that should ensue from the assumption, too many attempts have been made—not unnaturally by those who hold this faith—to adjust the evidence to the assumption. The result has been strained interpretations framed to meet one point, but without sufficient regard for the others. We repeat that we would not for a moment contend that the forces of erosion, the modes of sedimentation, and the methods of motion, are not the same in *kind* as they have ever been, but we can never admit that they have always been the same in *degree*. The physical laws are permanent; but the effects are conditional and changing, in accordance with the conditions under which the law is exhibited.

Such are the barriers which seem to us seriously to

retard the advance in one direction of an important branch of theoretical geology, whilst in another it is fronted by the stern rules of an apparently definite calculation.

We must ask to be forgiven if we cannot accept as conclusive the conclusions of Physicists respecting the extreme rigidity of the earth and the immobility of the crust. That the rigidity is now very great—as great, we will admit for argument's sake, as if the globe were of glass or steel—may be as it is asserted, but that conclusion can only be accepted in so far as it conforms to the facts of geology.[1] Were the data on which the conclusion is based fixed and positive, like those on which the laws of gravitation and light are established, there would be nothing for the Geologist to do but to bow to the decision of the Physicist, and revise his work. But in this case the tidal observations, on which the calculations of rigidity are mainly based, are of such extreme delicacy that, failing as the hypothesis does to satisfy the requirements of geology, the Geologist may be excused for his dissent, pending further inquiry. Should this tend to confirm the extreme rigidity of the globe, we must seek for some explanation of earth movements consistent with that rigidity. It is indisputable that up to the latest geological period—that touching on our own times—the mobility of the crust was very considerable, for the Raised Beaches of Europe and of the Mediterranean prove conclusively that in that period extensive areas were raised at intervals to heights of from 10 to 600 feet or more

[1] See Paper No. V.

above their former levels. It is difficult to conceive that a globe, of which the crust was then so mobile, could have acquired, in the comparatively short interval between the latest of the *beaches* and our own time, so great a rigidity as to be practically immobile.

For similar reasons the other opinion that the crust of the Earth is not less than from 800 to 2,500 miles thick is greatly open to question. We cannot imagine that a crust of that enormous thickness could, in such recent geological times, have possessed so great a flexibility as is indicated by the movements we have referred to. Independently of that improbability, there are certain geological facts which are inexplicable on that assumption. Volcanic phenomena would be unintelligible; for vents traversing that thickness of solid rock could hardly be kept open owing to the cooling which the lava in its ascent would undergo.[1] The rock-fragments ejected during explosions are also those of rocks which lie at no great depth, while, with the increase of temperature in descending beneath the surface,[2] there is every reason to suppose that at a depth to be measured by tens, and not by hundreds of miles, the immediate underlying magma at least is in a state of plasticity such as would allow of comparatively free movements of the crust. Again, surely, if

[1] See Paper No. IV. It has been suggested that lakes of molten lava may exist at moderate depths in the thick solid crust. But this is a mere assumption. Amongst other objections to it is that, if that were the case, as the lakes were discharged the overlying crust would sink, and volcanic areas would always be areas of subsidence, which is contrary to fact.

[2] See Paper No. VI.

the crust were so thick, we might expect to find, when that crust was broken and its edges thrust up by compression or protrusion of the igneous rocks, that some indications of that enormous thickness should be exhibited; but none such are forthcoming. Whatever may be the state of the nucleus, there is nothing geologically to indicate, as many Physicists likewise have contended on other grounds, that the outer crust of the earth is more than from about twenty to thirty miles thick. The effective rigidity will therefore, if it be necessary, have to be explained in some other manner than that of a comparatively solid globe or of a crust of enormous thickness.

We are thus brought face to face with apparently irreconcilable opinions. That they admit of adjustment there can be no doubt, but it must be by mutual understanding. How it is to be effected is a problem for the future.

These, briefly, are the barriers which restrict inquiry on many important questions. On the side of the Uniformitarians, it is assumed that every position must be reduced to a fixed measure—where fixity is not possible—of time and speed; and, on that of the Physicists, Geologists are gently reminded that the subject is outside their immediate sphere of inquiry, in a way somewhat suggestive of "*the closure.*"

It would be an unfortunate day for any science to have free discussion and inquiry barred by assumed postulates, and not by the ordinary rules of evidence as established by the facts, however divergent the conclusions to which those facts lead may be from the prevailing belief. In any case it must be remembered that no hypothesis can be true which does not satisfy the conditions both of the *geological phenomena* and of the *physical laws*.

The foregoing remarks are intended to apply mainly to questions connected with the more recent geological periods. The older epochs have happily been treated as beyond the barriers, and consequently Geologists have there enjoyed and made good use of their greater freedom. It is to be hoped that, when the phenomena of these later periods are judged of by the evidence of facts rather than by rules, they will receive more independent interpretations—interpretations that may escape the dwarfing influence of Uniformitarianism.

II

CONSIDERATIONS ON THE DATE, DURATION, AND CONDITIONS OF THE GLACIAL PERIOD, WITH REFERENCE TO THE ANTIQUITY OF MAN

WHEN, twenty-eight years ago, the barriers which restricted the age of Man to a limited traditional chronology were overthrown by the discoveries in the Valley of the Somme and in Brixham Cave, the pent-up current of geological opinion tended to the other extreme of assigning to Man (Post-glacial) an antiquity unwarranted by the facts. The belief in that great antiquity seems also to be exerting, in a manner somewhat similar to that which at first caused the rejection of the Kent's Cave and Somme Valley evidence, a perceptible bias on the questions raised as to an earlier age of Man, apart from the question of Pliocene or Miocene Man, into which, on the present occasion, I do not purpose to enter.

The extreme opinions which dealt with millions of years are now probably held by few; but still many and, probably, the majority of Geologists assign to the Glacial and Post-glacial periods, which involve the question of the antiquity of Man, vast periods of time. At the outset of the discussion in 1859, and when the antiquity of Man was limited to the Post-glacial period, I saw no reason for assigning to that period the length of time then claimed by many Geologists. The data, however, on which to form an opinion with respect to the duration of this and of the Glacial period were too imperfect to enable us to come to any definite conclusion. Since then further observations in the Alps, and more especially the observations of the Danish Geologists in Greenland, have brought forward facts of great importance in their bearing on ice-action and growth, and have furnished us with data which may warrant the estimates I now beg to lay before the society, not as a sufficient discussion of the subject, but as a preliminary inquiry.

Measured by our own limited experience, the excavation of the valleys, the life of the successive generations of the Pleistocene Mammalia, and the dying-out or extinction of a large number of species might seem to demand a long period of time. Consequently at first it was suggested that the Glacial period commenced possibly about a million years since, and that the Post-glacial period had lasted about 200,000 years. It was felt however, on the other hand, that the large proportion of existing species of Vertebrata and Invertebrata which came in with the Pleistocene period and had since undergone no change, combined with the stationary condition of Man himself during so long an interval, presented serious objections to adopting such lengthened periods of time. But on neither side were the conclusions based on any definite geological data. To the

Uniformitarian the assumption of limitless time was an indispensable need.

The question was in this state when Dr. Croll's attention was directed to the subject. After investigation of existing hypotheses he came to the conclusion that the cold of the Glacial period could not be ascribed to any physiographical changes in the distribution of land and water, but that changes in the eccentricity of the earth's orbit afforded a probable clue to great secular variations of climate, such as would produce Glacial epochs. Following up Leverrier's calculations, and assuming that the periods of greatest cold were when the eccentricity rose to a high value, and that the warmer periods occurred during the times of lesser eccentricity, Dr. Croll, by an elaborate mathematical computation, extended the inquiry as to the extent and periods of maximum and minimum eccentricity for 3,000,000 years back and 1,000,000 years to come, and showed that within the last million years there have been two such periods of extreme eccentricity—the one extending from 980,000 to about 720,000 years ago, and the other from about 240,000 down to 80,000 years ago.[1]

As the former period was of greater duration than the latter, and the eccentricity also then attained its highest value, Dr. Croll was at first disposed to refer the Glacial epoch proper to that period, and to consider the latter as corresponding with the extension of the local glaciers towards the close of the Glacial epoch. On this point he states that he "consulted

[1] *Climate and Time*, chap. xix. (1875).

several eminent Geologists, and they all agreed in referring the Glacial epoch to the former period," the reason assigned being that they considered the latter period to be much too recent and of too short duration to represent that epoch.

Dr. Croll therefore had fair warrant that he was well within the limits of geological probabilities when, on a reconsideration of the subject, he came to the conclusion that the Glacial epoch must be referred, not to the first-named period of eccentricity, but to the later one, commencing 240,000 years ago; and this is a date now very generally accepted. He considered that "the modern and philosophic doctrine of uniformity had led Geologists to overestimate the length of geological periods" (p. 325). Nevertheless he assumes that "the present rate of subaerial denudation does not differ greatly from that which has obtained since the close of the Glacial epoch" (p. 338), and proceeds with the argument in accordance with this view, taking the rate of denudation to be one foot of soil removed from the surface in 6,000 years—an estimate founded upon the quantity of sediment now carried down to the sea by such rivers as the Mississippi, the Rhine, the Rhone, the Ganges, &c. This is, in fact (notwithstanding the qualification just admitted), applying the results of experience in recent times to geological periods when the meteorological phenomena were evidently totally different, and the exact conditions unknown. Consequently the data cannot be applicable. If used at all, as a sort of base-line, it must be with modifications, such as taking existing data as the known quantity, and adding an unknown quantity "x," of which we can only judge by the character of the phenomena of the different periods. Some of the objections which have occurred to me against comparing the present rate of denudation with that of past times and of the Glacial period I have given elsewhere,[1] so need not repeat them. Nor is it my intention to discuss Dr. Croll's theory upon his own special grounds, which he has argued with much ingenuity and presented in so attractive a form that, as I am well aware, it has carried conviction to a large number of Geologists. That the causes he assigns have a certain value, there can, I think, be little doubt, and they may have an important bearing upon certain geological problems, such as relate to some periods of low temperatures which seem to have prevailed at certain geological periods, for example, that which in early Eocene times, affected the marine fauna of the Thanet Sands and the flora of Gelinden; but that they are sufficient to account for the extreme cold of the Glacial period is open to question. There are, as Dr. Croll himself observes, Astronomers and Physicists who are of opinion that the climate of the globe could never have been seriously affected by changes in the eccentricity of its orbit. The point has been contested by Professor Newcombe, the Rev. E. Hill, and others, whose papers and Dr. Croll's rejoinders should be consulted.[2]

If the cold were due to this cause, how is it that,

[1] *Geology*, vol. i. chap. vi.
[2] Various papers in the *American Journal of Science, Philosophical Magazine*, and *Geological Magazine*, from 1876 to 1884.

whilst the lesser eccentricity of 240,000—80,000 years ago resulted in the admitted Glacial epoch of the Pleistocene period, the effects of the greater and longer eccentricity of 980,000—720,000 years ago, which should surely have resulted in a still more intense Glacial period, has not left its traces in some anterior geological series. But there is no evidence of it in any Tertiary period.

It is, however, not only on the question of the sufficiency of the hypothesis to account for the facts, a point which I would leave to Astronomers and Physicists to decide, but it is on the geological question whether the necessary concordance exists between the observed phenomena and the phenomena as they should be were we to accept Dr. Croll's views, that we should judge of its applicability. If we adopt the hypothesis it should follow:—

> 1st. That at intervals during all geological time there would be certain periods during which a recurrence of similar glacial conditions took place.
>
> 2nd. That interglacial conditions would affect each pole alternately, and that there should be in both hemispheres warm interglacial periods.
>
> 3rd. That the commencement of the Glacial epoch should be placed about 240,000 years back, and have a duration of 160,000 years; after this the amelioration of the climate to its present condition, involves a lapse of time of about 80,000 years.
>
> 4th. That the date assigned to the disappearance of Palæolithic Man was not less than 80,000, whilst, if carried back to Pre-glacial times, it would necessitate an antiquity of nearly 300,000 years.

With regard to the first point, Dr. Croll shows that in the three million years for which his tables are computed there were five periods during which the eccentricity was as great as or greater than during the Glacial epoch proper; so that, taking geological time at the hundred million years, at which he estimates it, there should have been in all probability some 100 to 150 such periods of cold; but with the exception of the Permian, which is still under consideration, where is there evidence of any such cold periods?

Dr. Croll does not overlook this difficulty, and contends that in the Italian Alps there is strong evidence in favour of the opinion that glacial conditions existed there during the Miocene period. This, he informs us, is stated on the evidence of two distinguished Geologists; but the fact of the Miocene date of the beds in question has never since been confirmed; all the later evidence tends to show that it was not until towards the close of the Pliocene period that glacial conditions set in in the Alps. At the same time he admits the existence of warm conditions during Miocene times, which influenced even the flora of Greenland at that period.

For the Eocene period Dr. Croll relies on another Alpine case—the coarse conglomerates with its large blocks forming the Flysch. But this was a period of Alpine disturbance and change, when, though the rocks may have been rent and worn down in the mountain area, the marine life at a short distance gives evident indications of a high general temperature: Nummulites then abounded in the surrounding seas, together with Echinoderms of a decidedly tropical aspect.

The case for the Chalk is still weaker, for the very

few and exceptional foreign rock-boulders that have been found in it are of small size, such as might have been carried in the roots of trees or by seaweeds, or possibly by small winter ice-rafts from the coasts north of Scandinavia or others, whilst all the life of the Cretaceous sea itself is strictly that of temperate, if not of warm, latitudes.[1] The small pebbles found occasionally may have been carried by the large marine reptiles.

Facts of the same character as these Dr. Croll would have us accept as evidence of the action of ice in Scotland during the Oolitic period; but in face of the incompatible fact that at those times warm conditions of climate extended far North, and that corals, Cephalopods, and huge reptiles swarmed in the seas, this can hardly be admitted.

The climatal conditions during the Permian period may be open to doubt; but on this point it is not necessary here to enter. If a cold time be admitted, it would not affect the general question.

Nor is it easy to admit the claim for ice-action during Carboniferous times when the luxuriant vegetation of the Coal-measures flourished not only here but on Bear Island and other northern lands. With respect to the blocks of granite alluded to as occurring at the base of the Coal-measures in central France, they may be, like the Tors of Cornwall, blocks left *in situ* from the decomposition of the granite on which the

---

[1] It is possible that blocks carried down by glaciers of high mountain ranges may have been widely dispersed in the seas of various geological periods without involving the need of Glacial epochs.

Coal-measures there rest, or they may be boulders washed down at that period by the torrents from the adjacent granitic, and then higher, mountains. Other foreign pebbles may be accounted for as we have accounted for those in the Chalk (and last note).

There are similar palæontological objections to ice-action in the Devonian and Silurian periods. Although there may be at times instances in which the blocks show striæ and are derived from rocks not known in the locality, it must be borne in mind that such striated masses may be fragments of slickenside surfaces in the rocks from which the breccias are derived; and that, although a particular rock may no longer show in the locality, it may exist buried beneath newer deposits, as in the case of the granite of the Ardennes, which, although formerly unknown in a particular district, was met with in a railway-cutting beneath a slight covering of Palæozoic rocks.

Admitting the imperfection of the geological record, it is evident that, as a whole, the adduced instances of cold periods fail entirely to supply any sufficient evidence, either in their intensity or number, to support the theory of recurrent Glacial periods. Surely out of 150 or even 100 possible instances indicated by the theory, more definite evidence would be found, especially in the more recent periods, of which the strata are so accessible.

With respect to the second point, Dr. Croll states that "the Glacial epoch may be considered as contemporaneous in both hemispheres. But the epoch consisted of a succession of cold and warm periods, the cold periods of one hemisphere coinciding with the warm periods of the other, and *vice versâ*" (p. 234). This involves an indefinite succession of interglacial periods, but I do not see what evidence there is in Europe of such a succession of interglacial periods. Dr. Croll accounts for this on the ground that "the geological evidences of the cold periods remain in a remarkably perfect state, whilst the evidences of the warm periods have to a great extent disappeared" (p. 238). If, however, one instance is admitted, might we not expect other instances to have been preserved in some of the many localities affected? especially as he estimates the average duration of a warm period at about 10,000 years.

There are, it is true, indications both in this country and in Switzerland of intervals of milder conditions during certain times of the Glacial epoch; but these minor differences may be due, not to the cosmical cause assigned by Dr. Croll, but to those changes of climate that could be brought about by differences in the distribution of land and water, such, for example, as the extensive submergence which took place in England and Northern Europe, after the first great land glaciation, when there was a submergence in England probably of from 1200 to 1500 feet, and a large land-area became covered by the sea—a change which could hardly have failed to affect the climate of Switzerland and other parts of Europe.

The third and main point, namely, the probable duration of the Glacial epoch, hinges materially upon what we know of the growth of glaciers, and of the rate at which they move. Twenty years ago the only data we had bearing on the subject of glacier motion and action were the investigations of Agassiz, Forbes, Tyndall, and others on the glaciers of the Alps. These observers had determined with great care the rate of motion of many of the Alpine glaciers, their periods of advance, and the causes of their growth and decay; and it was on the data thus obtained that Geologists had to judge what might have been the rate at which the ice moved during Glacial times.

It was found that in July the Mer de Glace advanced at the rate of 33 inches in the twenty-four hours, and the Aletsch glacier in August 19 inches; whilst the winter rate was estimated at about half that of summer; so that the advance of the ice for the year might lie between 300 and 400 feet, though in one year the Mer de Glace was found to have advanced 483 feet.

The motion of Hugi's hut on the Aar glacier, which was based on the longer average of fourteen years, showed it to have been at the mean rate of 338 feet annually. We may therefore take the average motion of the Swiss glaciers to be about equal to 300 or 400 feet annually. To this there are certain exceptions.

In the cold summers of 1816 and 1817 there was a general advance of all the Swiss glaciers, whilst since 1856 there has been a general retreat. In the two years above named some of the glaciers advanced from 100 to 150 feet or more. Of other Swiss glaciers an advance of 1 mètre daily in summer is recorded; but the most remarkable case is that of the Vernagt glacier in the Tyrolese Alps, which is exceptional in that it advances by fits and starts. Between 1843 and 1847 the ice from this glacier covered the valley below to the length of 1264 mètres, which gives a mean annual rate of 870 feet, while the thickness of the mass of ice near its extremity exceeded 500 feet. Another glacier in the same range advanced, after many oscillations, above a mile in a century. On the other hand the Rhone glacier between 1856 and 1877 retreated nearly half a mile, or on an average 116 feet annually; whilst between 1870 and 1877 the retreat extended for a distance of 400 mètres, or a mean of 187 feet per annum, the greatest retreat being between the years 1870 and 1874, when it amounted to 250 mètres, or a mean of 205 feet per annum. In the Valley of Chamouni the Glacier du Tour retreated 320 mètres in the eleven years between 1854 and 1865, or at the rate of about 155 feet annually. The mean annual ablation of the Swiss glaciers is estimated at about 10 feet, but on the Glacier des Bossons the surface of the ice has been lowered 260 feet in twelve years, or nearly 22 feet annually.

These rates give a mean possible annual advance during cold seasons of from 200 to 300 feet. Taking for the present a mean rate of 250 feet, the glaciers of the Rhone, which during the prolonged cold of the Glacial period had a length of 250 miles, might have travelled that distance in about 5000 years. This, however, is assuming the absence of seasonal retardations and fluctuations, which is not possible. Allowances have to be made for warmer seasons and temporary retreats of the ice, and also for the fact that the old glacier did not move on the steep inclines of Alpine valleys, but traversed the small incline of a great river-plain. On the other hand, we have to take into account the circumstance that the present seasonal changes give no measure of the growth of ice under the continuous and more extreme glacial conditions of that Epoch. Estimates of the great length of the Glacial period based on these rates are very unreliable. I give figures merely for the sake of affording a parallel for comparison.

We have, however, in Arctic regions truer and more adequate terms of comparison with former rates in the great ice-sheet of Greenland. Already in 1876 Professor Helland[1] showed that the Greenland glaciers had a much more rapid rate of flow than those of Switzerland. The Jakobshavn glacier, notwithstanding its small slope of only half a degree, was found to advance its front at the rate of from 50 to 60 feet a day. The flow of the glacier of the Fjord of Torsukatak, which is nearly five miles broad, gave a rate of from 12 to 33 feet daily. Taking the average rate of the three glaciers on which Professor Helland made observations, the average discharge of the ice measured 23 feet in twenty-four hours; and he estimated that at the Jakobshavn glacier only four years would be required to transport a mass from the edge of the inland ice to the sea, a distance of $12\frac{1}{2}$ miles. But he considered it improbable that the inland ice would move with anything like the velocity of the glaciers, and calculated that the mass of ice starting halfway between the east and west coasts of Greenland would take eighty-one years to reach the fjord. These observations were, however, made in the summer months, and

[1] *Quart. Journ. Geol. Soc.*, vol. xxxiii. p. 142.

were only of a few days' duration, so that the annual rate was not determined.

Since then a Danish scientific expedition, consisting of Engineers and Geologists (one of whom, Mr. K. J. v. Steenstrup, passed eight summers and two winters in the country), have completed a most important exploration of the Greenland ice, of which a short summary has recently been given by Dr. Rink.[1] Their observations fully confirm those of Professor Helland, and show that the motion of the inland ice may be *compared to an inundation*.[2] It was found that there is a general movement of the whole mass of the ice from the central regions towards the sea, and that it concentrates its force upon comparatively few points in the most extraordinary degree. These points are represented by the ice-fjords, through which the annual surplus of ice is carried off and discharged in the shape of icebergs.

The velocity of the ice was noted in seventeen glaciers, the measurements being repeated during the coldest and the warmest seasons; and it was found, remarkable as it may seem, that the movement was not materially influenced by the seasons. The great glacier of the ice-fjord of Jakobshavn, which has a breadth of 4500 mètres, was rated at 50 feet per diem. One of the glaciers in the ice-fjord of Torsukatak has a movement of between 16 and 32 feet daily. The large Karajak glacier, about 7000 mètres broad, proceeds at the rate of from 22 to 38 feet in twenty-four hours; and another in the fjord of Jtivdliarsuk,

---

[1] *Trans. Edin. Geol. Soc.*, vol. v. p. 286 (1887).
[2] The italics are mine.—J. P.

5800 mètres broad, at from 24 to 46 feet. The conclusion at which the Danish party arrived was that the glaciers which produce the bergs move at the extraordinary rate of from 30 *to* 50 *feet per diem throughout the year.*

What, then, may have been the rate of movement of the great ice-sheets of America and Europe in the Glacial epoch? No doubt the velocity of the ice in the ice-fjords is increased by the free play of the ice as it reaches the sea and by the rapidity with which the bergs are detached. It is also increased by the circumstance that the great body of inland ice, the whole of which is in motion, is forced, and has to escape, through the passes between the range of mountains which fringe the coast and rise high above the districts immediately inland. The summits of those mountains form bare and isolated masses (Nunataks) standing above the surrounding ice-sheet; and the passes between them, and through which the ice forces its way, have been gradually worn down, and now form the channels through which the surplus inland ice escapes, with a velocity increased in proportion to the contraction of the passages. Thus in the ice-fjord of Torsukatak, which is nearly 5 miles wide, the ice passes out with a mean velocity of 24 feet per diem, or equal to a mass of ice of that width, and $1\frac{3}{4}$ mile long, annually; the Karajak glacier, which is $4\frac{1}{3}$ miles broad, flows at the rate of 30 feet daily, or equal to a length of above 2 miles a year; and in the huge ice-fjord of Jakobshavn, which is not quite 3 miles broad, the ice attains a velocity of 50 feet daily, so that a length of above 3 miles of ice is discharged annually.

Until all the glaciers have been gauged, and we know the relation of their totals to the breadths of the intervening "Nunataks," no definite measure of the total volume of annual surplus ice can be established; but for our general purpose some approximate idea may be formed. The average of these great glaciers gives a mean rate of 35 feet daily and a discharge of $2\frac{1}{5}$ miles of ice annually. There are other ice-fjords of far greater breadth than those, such as the Humboldt glacier, which is 60 miles broad. Looking at the map, it seems not improbable that the breadth of *ice-front* to *rock-front* of the whole coast may be in the proportion of 1 to 20.

Supposing the quantity which runs off to be equalized throughout the whole extent of coast, the fringe of ice which would pass off from the land annually would be $\frac{1}{8}$ of a mile long, or a length of one mile of the ice sheet would take eight years to pass off. If the proportion of ice to cliff should prove less, say 1:30, then it would take twelve years. In the one case a sheet of ice 100 miles long and of the width of the central axis would require 800, and in the other 1200 years for its discharge. Applying these measures to estimate approximately the time-rate of the old continental ice sheets, such as those which originated in the North American highlands and the Scandinavian mountains, and taking the length of their extension from their culminating point at, say, 500 miles, the time required to form this length of ice would be respectively 4000 and 6000 years.

This, however, is based only on one known quantity —that of the mean rate of the Greenland ice-sheet. We have, on the *per contra* side, to allow for certain unknown quantities. First, allowance has to be made for the difference between the free escape into the sea and the impeded progress of ice over land with slight gradients. On the other hand the great thickness and weight of the ice in the central area, where in the Glacial epoch it attained a thickness of from 5000 to 6000 feet, have to be taken into account.[1]

The mass of ice, projected outwards towards its circumference, might, except where it met with contracted channels, roll over the land as a plastic body with comparatively little friction. When, in the Glacial epoch, the great southern glaciers of the Alps flowed down the steep and confined valleys opening upon the flat plain of Lombardy, they deeply ploughed their channels, and pushed before them for short distances enormous moraines; but in the wide open tracts of the United States, of Northern Europe, and central England, where the ice met with little resistance and could expand in several directions, there is, as a rule, an absence of moraines and often of glacial striæ.

In the second place, there were, no doubt, seasonal fluctuations which would retard the flow for lesser or greater periods. It is asserted that in Europe there were interglacial periods during which the ice disappeared from the surface for great lengths of time. But either the evidence is insufficient or it points to

[1] The American Geologists also consider that the Canadian land then stood considerably higher than at present.

slight temporary effects, except in one case, which is of more importance, and on which great stress has been laid, namely, that of Dürnten in Switzerland. Beds of lignite with mammalian remains are there intercalated between two glacial deposits. Admitting the fact that the lignite rests on beds of undoubted glacial (ground-moraine) origin, and that the trees grew on the spot where their stumps and remains are found, it by no means follows, as contended, that because these trees are all of species now living in Switzerland the temperature was that of Switzerland at the present day. *Pinus sylvestris, Abies excelsa,* the Yew, the Birch, and the Oak flourish equally in Sweden and far north in Siberia. On the other hand, there is one species of *Pinus* (*P. montana*) which is spread over the mountain districts up to heights of 7000 feet, and is rare in the low lands; while one of the mosses is closely allied to a species now growing on the hills of Lapland. The few species of Mammalia have a distinctly northern facies. *Elephas primigenius, E. antiquus, Ursus spelœus,* while *Cervus elaphus* and *Bos primigenius* are commonly associated with the Reindeer, Musk-ox, and other Arctic animals. Further, both the trees and animals are of the same species that are found in our " Forest-bed "—the last land-survival before the setting in of the extreme glacial cold, and indicating a climate of not inconsiderable severity.

Should the return, therefore, of the glacier, after its first retreat, or the inset of colder conditions, be ascribed to anything more than a comparatively slight temporary change of climate, like those that now for a succession of seasons cause, from time to time, a temporary advance of the glaciers, only more marked, though we may allow for greater differences and longer intervals of time than now obtain?

Such minor vicissitudes of climate are more compatible with changes in the physiography of Europe than with the cosmical causes to which the Glacial epoch, as a whole, was in all probability due. Nor is it difficult to find such a cause in the extensive changes in the distribution of land and water which took place in Britain and Northern Europe after the first great land-glaciation and the formation of the Lower Boulder-clay. The submergence of Ireland, Wales, Scotland, and England (in part), and of a large area in Russia and North Germany, extending to Holland, was sufficient, with the influence of currents from the south (for in the shells of the Middle Boulder-series there is a large percentage of southern forms and an absence of extreme Arctic forms), to effect a considerable amelioration of the climate, such as would lead to the temporary reappearance of a less northern fauna and flora.

With the rise of the temporarily submerged lands the climate again changed, and brought back colder conditions and winter-snow floods which overwhelmed in their course the forest-growth that had sprung up in the meantime, and covered it with beds of stratified sand and gravel.[1]

[1] The angular boulders which lie on the surface of the fluviatile sands and gravel are supposed to have been brought down by the returning glacier. But how is it that the glacier, which, in the first instance, ground down and bared the rocks, failed to plough the loose fluviatile beds in its last advance?

For the formation of this lignite deposit a period of 6000 years has been claimed; but the claim rests on very doubtful data. The bed varies from 5 to 10, and is rarely 12 feet thick. In the estimate the maximum thickness of 12 feet is taken, and it is assumed that to form this 12 feet of lignite it would have required 60 feet of peaty matter,[1] or that it took 5 feet of peat to form 1 foot of lignite, and that 100 years would be needed for the growth of each foot of peat: thus a total of 6000 years is obtained. But the growth of peat varies extremely. It may be, in some cases, not more than 1 foot in a century, but it is commonly more, being sometimes as much as 4, 5, and even 10 feet in that time; and while it is estimated that to form 1 foot of coal, from $2\frac{1}{2}$ to 3 feet of woody matter may be required, it is clear that lignite, which has lost less of its original constituents[2] than coal, and of which the specific gravity is about 1·25, while that of coal is about 1·5, cannot require for each foot 5 feet of peat and wood. Taking, therefore, the original thickness of the bed at 24 instead of 60 feet, and the growth at 2 to 4 ft. in a century, 600 years, or at the outside

---

[1] It is formed more of wood than of peat.

[2] In the extreme case of the conversion of wood and peat into anthracite, in which the proportion of oxygen and hydrogen to the carbon is as 5 : 95, the estimate is of from 7 to 8 feet of wood to 1 foot of anthracite; and in ordinary coal, where these constituents are roughly as 15 : 85, the estimates vary from $2\frac{1}{2}$ to 3 feet (in some estimates as much as 5) of woody matter to 1 foot of coal. In lignite, then, where the change has involved less loss (say to 30 O + H : 70 C) and the pressure has been less, the compression must certainly have also been less.

1000 years, instead of 6000, might have sufficed for the formation of the Dürnten beds.

Although, therefore, these breaks may involve considerations respecting hundreds, they are scarcely likely, as they must have been subordinate to the general advance of the ice-sheet, to involve questions relating to thousands of years. It is to be observed also that there is no evidence in North America of an interglacial period in the sense of the one supposed to have existed in Europe, although there is evidence that after the great ice-sheet had retreated for a very considerable distance northward, there was a pause followed by a partial advance again—an advance marked this time by deeply lobed lines of moraines.

While, however, admitting that these breaks prolonged the duration of the Glacial epoch, there are other factors in the question which may have tended to shorten it. At present the discharge of ice from the Greenland sheet is merely the annual surplus under conditions of a settled mean temperature; but the Glacial epoch was a time, on the whole—although there may have been breaks—of constantly increasing cold, and of constant increase in the area of the great ice-sheet, and therefore there was not merely a supply due to a uniform mean annual temperature, but the increments arising from the gradual secular refrigeration.

It may also be a question whether the rainfall was not then greater than now. At present in Greenland it appears to be small, but in the North American old ice-area it amounts to 40 or 45 inches annually. The Florida promontory, which now deflects and contracts the Gulf Stream, was at that early period considerably smaller, as the coral reef by which it is formed had not then extended so far south. Consequently, the channel through which the stream passed being wider, a greater volume of warm water could have flowed through; and this, passing into the North Atlantic, would have materially affected the precipitation both in North-eastern America and North-western Europe.

The growth of the ice-sheet is not, moreover, dependent only on the rainfall. The experiments of MM. Dufour and Forel have shown that when the temperature of the air on the Rhone glacier varied from 41° to 52° F. there was a condensation of moisture equal to 150 cubic mètres of water per square kilomètre, and this increased proportionally the volume of the glacier-ice and water. Owing to these and other circumstances, such as ablation, a difference of level in the surface of some of the Swiss glaciers to the extent of from 80 to 100 feet has been known to have been effected in the course of twenty years.

The observations on the Greenland glaciers have now shown us that the continental ice advances at the rate of one in eight or else in twelve years, so that an advance equal to 500 miles (the span of the old ice-sheet) might be made in 4000 or 6000 years. Allowing for differences of conditions and seasonal checks, and, on the other hand, taking into consideration the cumulative effects of a persistent and constantly increasing cold, these rates afford grounds for assuming a far more rapid spread of the old ice-sheet than the rate of motion of the Alpine ice could possibly suggest. Taking the vast difference of the two rates, a term possibly of 20,000 to 25,000 years might have sufficed for the extension of the old ice-sheets, but no measure can at present be otherwise than approximate.[1]

In any case if before the important observations in Greenland, it was held that the Alpine-glacier mean annual rate of 250 feet was consistent with the age then ascribed to the Glacial period, surely the mean annual rate in Greenland of 14,600 feet can only be consistent with the much shorter term in which the same amount of ice-work could be done.

The acceptance of the longer periods of Croll was very much the result of the belief that no shorter time would account for the excavation of the valleys supposed to have been formed during the Post-glacial period. There can, however, be no doubt that the great valley-erosions took place in Pre-glacial and Glacial times; and that that of Post-glacial times was comparatively slight. Anyhow these differences must be taken into account in considering the length of duration of the Glacial Period.[2]

The adoption of a rate of denudation based on that of the present day has always seemed to me open to grave objections, and in this belief all

---

[1] At all events it is obvious that if a term of 200,000 to 300,000 years was held as being in accordance with the Alpine rate, it cannot be in accordance with the Greenland rate. Either one is too short or the other too long.

[2] Though little change has yet been made in the line of argument, there has been a growing belief amongst Geologists that the present rate of change has not always been uniform, and must not be taken as the measure for all past and all future time.

subsequent experience confirms me. Dr. Croll, who, with others, adopted the generally accepted rate of denudation, namely, one foot of rock or soil removed off the general level of the country during 6,000 years, nevertheless remarks "if the rate of denudation be at present so great, what must it have been during the Glacial epoch? It must have been something enormous."[1]

It is no more possible to judge of the rate of denudation during the Glacial period by that of the river-action at the present day than it was to estimate the rate of flow of the Greenland ice by Alpine experience. The enormous pressure and friction exercised by glaciers from 1,000 to 6,000 feet thick in contracted valley-channels, especially in fjords, where, as for example in Norway, it stood in Glacial times from 1,800 to 2,000 feet higher than now; the powerful disintegrating effects of extreme cold on rocks; the annual action of ground-ice in rivers, and of the sweeping and devastating floods, resulting from the melting of the winter's snow and surplus ice, combined to produce results of which it is impossible to judge by the ordinary work of these temperate latitudes. We must go to Greenland to find terms of comparison.

I need not at present go more fully into this subject, but I would just allude to some interesting corroborative testimony recently brought forward by Prof. J. D. Dana in connexion with the phenomena of the Connecticut valley.[2] The numerous old river-terraces in this valley extend for a distance of 250

---

[1] See Paper No. 1, p. 9.
[2] *American Journ. of Science* for March 1882.

miles inland. The river has excavated a valley through the ancient high-level plain to a depth of from 150 to 200 feet, with a width of from one eighth of a mile to one mile. The mean depth of this river in flood at the Postglacial (Champlain) period is estimated by Dana to have been about 140 feet, the mean height of the present floods being about 26 feet. The mean width of the upper section of the flooded stream he estimates to have been 6,000 feet. Taking these measures, together with the mean slope, he obtains a maximum velocity of over twelve miles an hour, with a mean of about three or four miles, whence some idea may be formed of the enormous transporting power of the river of that period. The annual rainfall in this district now varies from 65 inches on the coast to 42 inches in the interior; but during the Glacial period, Dana considers that the special conditions must have occasioned a much more abundant precipitation— possibly as high as 120 inches.

In this country and in the north of France the valleys do not seem to have been excavated to the depth of more than from 80 to 120 feet in Postglacial times. It may be difficult from our present experience to conceive this to have been effected in a comparatively short geological time; yet it is equally, and to my mind more, difficult to suppose that Man could have existed say 10,000 years (or 300,000, if Preglacial), and that existing forms of our fauna and flora should have survived during those long years without change and modification. The acceptance of Croll's dates, which would place the land-glaciation some 100,000 to 200,000 years back, would also lead to the difficulty (even on the assumption of a rate of denudation of 1 foot in 6,000 years)[1] that the surface-wear should have been far greater than it is. For example, to mention only two points ; (1) could the stria on surfaces of the rocks, which are acted upon by carbonic acid, have remained so sharp as they are, and (2) would not the limestone-rock, on which the boulders of Silurian rocks were left on the melting of the ice on the Yorkshire hills, show much greater wear than it actually does? These boulders now stand on pedestals raised from 1 to 2 feet above the surrounding surface-level in consequence of the dissolving away of the limestone rocks. We should look for pedestals of much greater height if the glaciation took place at the distant period involved in Dr. Croll's hypothesis.[2]

My original impression with respect to the Valley of the Somme was, that the high-level gravels originated in later Glacial times; that the intermediate stages and terraces were formed during the excavation of the valley as a resultant of the late-Glacial and Postglacial floods; and that the low-level gravels formed the concluding stage of those conditions. But since then the whole series has

[1] A general rate of this description is also scarcely applicable to a special rate, such as that relating to valley-denudation.

[2] In an old limestone quarry in the neighbourhood of Aix-les-Bains, surfaces left in Roman times have been since eaten into to the depth of a little more and less than an inch, which would give roughly a term of 20,000 to 30,000 years to account for the amount of 1 to 2 feet of the rock, calculated even at the present rate of wear.

generally been considered as Postglacial. The older date assigned would also agree better with the classification, based mainly on archæological considerations of M. Mortillet, though I should not place, as he does, these beds at the beginning of the Quaternary period. Possibly, however, our base line differs.

Much evidence has also since been brought forward with respect to the so-called Preglacial Man in England. The cave-work of Mr. Tiddeman and Dr. Hicks gives presumptive evidence of the early geological appearance of Man in the British area; and I see no reason to question the sub-boulder-clay evidence of Mr. Skertchly, although I was not successful in finding any specimens myself. Of the correctness of his opinion in respect to the stratigraphical position of the bed in which his specimen was found, I have, however, no doubt, and that it belongs to the Boulder Clay series. The great masses of gravel in the neighbourhood of Mildenhall and Lakenheath, likewise containing flint implements, are certainly not of fluviatile origin; they seem to me to be part of the phenomena connected with the great Boulder Clay system of the eastern counties, and belonging to the Glacial times.[1]

I am therefore led to conclude that the high-level beds of the Somme Valley at Amiens, of the Seine in the neighbourhood of Paris, of the Thames in places, and of the Avon at Salisbury, together with certain caves, date back to Glacial times, though not to its earliest stages. But while I cannot accept the antiquity of Man based on the prevalent Postglacial

[1] See Paper No. III.

time-basis, I would extend that antiquity on the evidence furnished by other drift deposits, and their relation to certain Pliocene beds. I need not, however, enter into particulars here, as the evidence in favour of early Glacial Man, is given in the next paper (No. III.)

In supposing that Man was present in this part of Europe in Glacial times, I am, however, far from claiming for him the antiquity which a term of 80,000 years would give to Postglacial Man, as usually understood. For the reasons before given, I believe that the Glacial epoch—that is to say the epoch of extreme cold—may come within the limits of from 20,000 to 25,000 years, and that of the so-called Postglacial period, or of the melting away of the ice-sheet, to within from 8,000 to 10,000 years. In any case, I believe that Palæolithic Man came down to within 10,000 to 12,000 years of our own times. If we may be allowed to form a rough approximate limit on data yet very insufficient and subject to correction, this might give to Palæolithic Man, supposing him to be of early Glacial age, no greater antiquity than perhaps about from 38,000 to 47,000 years; while, should he be restricted to the so-called Postglacial period, his antiquity need not go further back than from 18,000 to 27,000 years before the time of Neolithic Man in Europe.

Looking at the facts that so many of the species of our existing land and marine Fauna and Flora appeared in true Preglacial times, that is to say, in the time of the Forest-bed; that the great extinct Mammalia of later Glacial times have left no descendants, but have merely died out as a consequence of the great changes of climate and conditions; and the difficulty of conceiving that man could have existed for a period, say, of 100,000 to 200,000 years without change and without progress—looking, I say, at these facts, it seems to me that a shorter estimate of time is the only one in accordance with all the conditions of the problem. I submit that neither on the grounds of the time required for river-erosion, nor of the rate of glacier progress, can the demand for the extreme lengthened periods be maintained.

This view of the question also brings the geological and ethnological data into closer relationship. Palæolithic Man in North-western Europe disappeared with the valley-gravels. With the alluvial and peat-beds Neolithic Man appeared. In Europe we are unable to carry back historic Man beyond a period of from 3,000 to 4,000 years B.C. But already in Egypt, and now in Asia, it is proved that civilized communities and large States flourished before that time. Civilized Man must therefore have had a far higher antiquity in those countries, and probably in Southern Asia, than these 6,000 years. It is possible that the two periods may have overlapped, and that while Man in a more advanced state flourished in the East, he may here in the West have been in one of his later Postglacial stages.[1]

P.S.—The Greenland glaciers have shown us how rapid can be the rate of advance of an ice-sheet, but

[1] In the course of the discussion on this paper, stress was laid on the fact that we had no data for the *exact* measurement

they fail to give us any idea of what might be a rate of retreat. Nor do the Alpine observations furnish us with more than scant data. We find, however, in a paper by Mr. J. C. Russell, just published,[1] the needed information. It appears that in Alaska "the glaciers are slowly retreating, and probably have been retreating for 100 or 150 years. The amount of this recession in the case of the glaciers at the head of Yakutat Bay is known to be four or five miles, and at the head of Glacier Bay the retreat is thought to have been not less than fifteen miles during the past century." Here then we have evidence to indicate that the ice-sheet which covered the British Isles (taking it at 800 × 800 miles) could, as the recession must have gone on at all points of the circumference, have been removed within the limits, roughly speaking, of 3,000 to 8,000 years. Not that I would venture to assert that any exact value can be attached to those figures, but they show, at all events, how unsafe it is to appeal to the vast periods of time that have been claimed as necessary to effect these great revolutions of glacial times.

of geological time, in which I quite agree. Nevertheless, those who object adopt without hesitation the far more definite time estimates of Croll. I trust the data I have given, however imperfect, may serve to show that that estimate requires revision. My own estimate I submit merely as a nearer probable approximation to the truth. I do not pretend it to be an exact measure, though I am satisfied that the variation from the truth will prove to be but a fraction of the error I take to exist in Croll's estimate as adopted by those Geologists who base their opinion upon Uniformity of action and permanent slow changes.

[1] The *Scottish Geographical Magazine* for August 1894, p. 393.

III

ON THE PRIMITIVE CHARACTERS OF THE FLINT IMPLEMENTS OF THE CHALK PLATEAU OF KENT, WITH REFERENCE TO THE QUESTION OF AGE AND MAKE

1. Position of the Plateau Implements

It was in 1869 that Dr. (now Sir) John Evans, in company with the author and a party from High Elms, found on the chalk plateau at Currie Farm, near Halstead, a roughly made ovoid palæolithic flint implement.[1] The spot was 600 feet above the sea-level, and far from any river valley, though at no great distance from the head of the dry upper valley of the Cray, and within one mile of the edge of the chalk escarpment. Although we made further search over the field on the surface of which it was discovered, we did not succeed in finding any other specimen. There was nothing particular about the specimen, which might pass for a poor example of the ordinary river-valley type; nor was there anything in its surroundings to give definite clue to its geological age. But it was not until Mr. B. Harrison, of Ightham, began his persevering research on the

---

[1] *Ancient Stone Implements*, p. 531, and *Quart. Journ. Geol. Soc.*, vol. xlv. p. 295. It is by mistake that the specimen is stated to have been found on the second visit. It was on the first.

Eastern Plateau, between the Darent and Medway valleys, in the summer of 1885, that their numbers and peculiar forms have become known. A total of 1,452 [1] specimens have now been found. They occur scattered, at a large number of places, on the surface of the plateau at heights of from 400 to 800 feet, and extend to the very edge of the escarpment. Nevertheless, these rude implements are at first not at all easy to find, owing to their colour and rudeness of form. On the Western Plateau these implements have been traced by Mr. De B. Crawshay for a distance of nearly twenty miles between the Darent and Caterham valleys.

It is on the high ground only that the plateau implements occur, and though the heights actually vary considerably, this arises from the circumstance that the chalk plateau forms an inclined plane, having its highest pitch of 700 to 800 feet above the sea-level, along the line of the chalk escarpment, and thence falling by a gradual incline northwards to the height of about 400 feet. (Fig. 1.)

The following is a list of the localities at which they have been found, with the height above the sea-level, and the number of implements obtained at each; but fresh places are being constantly discovered.[2] In every instance, it is on the high levels of the places named that the implements occur.

[1] Mr. Harrison's collection now (September, 1894) contains 3,480 plateau specimens.

[2] There is little doubt that further research will show them to have a much wider range over the Chalk Downs and at other corresponding levels. Within the last two or three years some

The Plateau east of the Darent Valley

The implements have there been found at the different places named in the following proportion:—

Name of Place.	Height of Ground above Sea-level.	Number of Specimens.
	feet.	
Punish Farm	600	1
Hodsell Street	550	3
Sparksfield	520	95
Fairseat and Plot Farm	690	12
Wrotham Hill	760	15
Plaxdale Green, near Stansted	630	8
Stansted	633	19
Ash	490	235
South Ash	520	262
West Yoke Farm	460	222
Kingsdown	550	30
Terry's Lodge	770	50
Peckham Wood Corner	637	40
Gabriel's Spring Wood and Speed Gate	450	3
Horton Wood (west of)	400	3
Speed Plain and Gate	420	17
The Vigo	690	5
Birches Wood, St. Clere Hill	760	23
Cotman's Ash	665	4
Wick Farm	697	10
Bower Lane	520	88
Preston Hill (Shoreham)	510	67
Total	—	1,212

thirty to forty worked flints of the plateau type and colour have been found by Mr. Hilton, of East Dean, at Friston, near Eastbourne, 350 feet above the sea-level. Amongst them are specimens of the characteristic crescent-shaped scrapers. Mr. Crawshay has also found a few specimens at Stede Hill, above Lenham, east of the Medway; and the Rev. R. Ashington Bullen has found a well-characterised beak-shaped implement from Blean Hill, near Canterbury; and others from high ground near Amersham, and Micheldever, Hants; and from the neighbourhood of Axminster. More recently an important find has been made by Dr. H. P. Blackmore on a hill near Salisbury. Mr. A. M. Bell also reports a new locality on Bucklesbury Common, Berks. The specimens from Caddington, found by Mr. Worthington Smith on the chalk hills at Caddington, seem to be of a newer palæolithic type.

The Plateau west of the Darent Valley

The following is a list of specimens found on the West Plateau by Mr. De B. Crawshay and other observers:—

Name of Place.	Height of Ground above Sea-level.	Number of Specimens.
	feet.	
Cockerhurst (Shoreham)	450	3
Park Gate (Lullingstone)	430	1
Shacklands Wood (west of)	530	40
Hewitt's Farm (Chelsfield)	470	3
Oldmen Wood ,,	430	1
Polhill Plain [1]	447	22
Halstead Fields, north of Church	495	36
Morant's Court Hill	700	10
Colegates	500	1
Currie Farm	590	3
Norstead Hills	485	54
Botley Hill	875	5
Betsom Hill (Westerham)	790	23
Titsey Hill (Limpsfield)	864	11
Tatsfield Firs	820	3
Ivy Cottage (Tatsfield)	790	13
Park Wood	780	1
Farthing Street (Downe)	400	6
Total	—	236

2. The Geological Question

Upon the geological question relating to the age of the plateau drift it is not necessary to enlarge, as I have treated it in detail in two papers read before the Geological Society of London.[2] My object now is to inquire whether the character of the implements is in accordance with the early Glacial or Preglacial

---

[1] Mr. Bullen has recently discovered other localities to the north and south of the London Road, not far from the Polhill Arms, and within half a mile from the field on Currie Farm; and others on the east slope of Well Hill.

[2] Quart. Journ. Geol. Soc. for May, 1889, and for May, 1891.

age, to which, pending further inquiry, I would assign them.

I may, however, for the information of those members who are not acquainted with the geological argument, briefly give the facts on which the antiquity of the drift, with which the implements are associated, is established, in a diagram which embodies the essential points. (Fig. 1.)

FIG. 1.—DIAGRAM SHOWING THE RELATIVE POSITIONS OF THE IMPLEMENT-BEARING DRIFT ON THE CHALK PLATEAUS AND OF THE VALLEY DRIFTS OF POSTGLACIAL AGE.

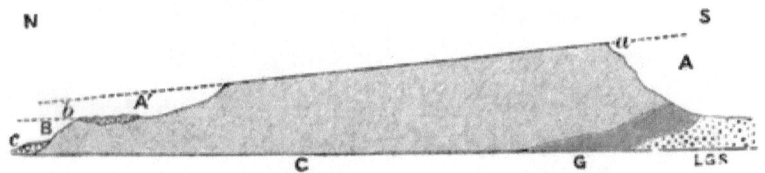

a.—Red Clay-drift, 5 to 20 feet thick, containing numerous unworn chalk flints, with some Tertiary flint pebbles, and with, here and there, remnants of Lower Eocene strata. A thin drift with Rude Flint Implements is scattered in places on the surface of the plateau.
b.—Valley gravel of fluviatile origin and late Glacial or Postglacial age, with Palæolithic Flint Implement and a few Mammalian remains. This lies about 100 feet above the level of the Thames.
c.—Low-level Valley gravel and loam, with Palæolithic Implements and numerous remains of large Extinct Mammalia. These slope down to the Thames level.
C.—Chalk. G, Upper Greensand and Gault. LGS.—Lower Greensand.

The chalk plateau rises from 400 or 500 to 600 or 800 feet above the sea-level, and is bounded north and south by the valleys A′, A. These valleys have cut off abruptly, on both sides of the plateau, the drift *a a* and some Pliocene beds, which originally extended over the area now occupied by those valleys, as shown by the dotted lines in Fig. 1. Spread in places on the surface of the plateau is a scattering of a peculiar drift consisting of water-worn brown-stained Flints, together with fragments of Chert and Ragstone,[1] derived from the Lower Greensand strata, which form a range of hills four to five miles south of the chalk escarpment. But at the time however when that *débris* was transported on to the chalk plateau, the valley A was still bridged over by the chalk and overlying strata which have since been removed by denudation. The plateau drift dates therefore from a time subsequent to the Pliocene period, though still when the beds of that age had a wider range southward, and anterior to the Glacial period, during which the valleys A′, A were doubtlessly excavated. This drift must be therefore of Preglacial or early Glacial age; and as the flint implements are closely associated with the plateau drift, and are limited to the area over which it extends, we are led to infer the Preglacial or early Glacial age of the men by whom they were fabricated.

The bed of the valleys A′, A afterwards formed the channels of the rivers that deposited the gravel *b*, and it is in these valley gravels, formed during the later Glacial or early stages of the Postglacial rivers, that the well-known Palæolithic flint implements of an advanced type—such as those of the valleys of the Thames and Somme—are found. As the rivers continued to deepen their channels, the gravel beds of this first stage were cut through and often left as terraces at various heights above the newer valley B, while in the bed of this later valley were deposited the gravels and loams *c* which constitute the low-level

---

[1] The "Southern Drift" of the Author (see *Quart. Journ. Geol. Soc.*, vol. xlvi. p. 155).

valley drifts of the last stages of the Postglacial rivers, and it is in these especially that the remains of the great extinct mammalia (the Mammoth, Woolly Rhinoceros, &c.) abound, associated with Flint Implements of later Man, though of types very similar to those of the higher-level valley drift b.

It will be obvious from the above, that the drift-beds a, with their associated implements, preceded those marked b, just as these latter preceded those of c. This establishes a *prima facie* presumption of the greater antiquity of the implements found on the chalk plateau—a presumption which is materially strengthened by the circumstance that the rude workmanship of the implements is in accordance with the great difference in geological age between the plateau and the valley drifts.

Nevertheless, though we may be warranted in drawing this inference from the facts so far as they are known to us, we yet need, as I have elsewhere explained (*Quart. Journ. Geol. Soc.*, vol. xlvii. p. 288), further information as to the exact relation of the implements to the plateau beds. Owing to the absence of pits and rarity of sections, we have had for the greater part to depend on specimens found on the surface, or thrown up in shallow plough furrows or trenches, or on a few roadside cuttings, and although from the deep staining of the implements, and their occasional *incrustations with iron oxide*, we have reason to believe that they have been embedded in a deposit beneath the surface, it is only in a few rare instances that they have actually been found at any depth. A large specimen was found at South

Ash in making a hole two feet deep for planting a tree, but as it was picked up on the thrown-out soil, its exact position beneath the surface remains of course uncertain. It was the same with the one obtained in a post-hole at Kingsdown. For two others we have, however, the personal testimony of Mr. Harrison. One he took out of a bank of red clay[1] on the side of a pond and at the depth of two and a half feet, and the other from a bed of "deep red clay" two feet in depth, at the Vigo.

The condition of the implements and their limitation to the surface of the plateau form is, however, evident, notwithstanding the rarity of definite sections. It is also clear from the occurrence of the implements on the highest summits of the chalk plateau, that whatever may be the causes which led to their distribution, those causes must have been, unlike these affecting the Valley Drifts and Implements, independent of the present topographical features of the district and of the system of river drainage as now established.

3. Alternative Explanation of their Origin

Is it possible, however, to account for the presence of these flint implements in any other way than that of their being contemporary or nearly so with the Southern drift? For example, could these implements, like the ordinary Neolithic implements which occur on the same ground, have been dropped on the surface where they are now found at some later date? The

---

[1] The Red clay found with some implements is, I believe, due to weathering of the "Red Clay with Flints"

answer to this is, that on these Neolithic implements there is absence of all staining—the only change being a loss of colour, the result of weathering by exposure on the surface. These also are found at all levels, whereas the plateau implements, besides their wear and colour, present all the physical characters due to having been *embedded in a special* drift, and *are confined to a special area*. The two sorts, although found on the same ground, remain perfectly distinct.[1]

Then again, is it not possible that similar rude specimens occur in the valley drifts, and have been overlooked owing to the prevalence of the better finished implements to which attention had been exclusively given? This, if we admit that rudeness of form alone is not a sufficient reason, is a fair argument. The ruder valley forms have not, however, been altogether overlooked. A large number of rude and badly finished specimens have been collected in the valley drifts, but they all belong to one set of types, and though I have seen and handled many hundreds of these, I question whether, with the exception of the *derived* specimens to be named presently, there were any like the ruder and most primitive of the plateau types.[2]

M. Boucher de Perthes collected everything in the Somme district which showed any traces of workmanship, howsoever indistinct, or even of similitude, yet I do not remember that in his great collection there

---

[1] For further observations on this point, see *Quart. Journ. Geol. Soc.* vol. xlvii. p. 133.

[2] There are, however, a few generalised types which seem to pass from the Plateau to the Valley epoch, and even to Neolithic times.

were specimens of the peculiar character of these plateau implements.[1] In Mr. Harrison's collection from the Shode Valley and the Ightham district there is nothing to correspond with the group of the plateau implements. They are all of the valley type. At my request he has re-examined several of these localities, as well as the large pit at Aylesford in the Medway Valley, and the pits at Milton Street (Swanscombe) in the Thames Valley, with this special object in view. He reports to me that he finds no contemporary specimens of the plateau type, and very few derived specimens of that type either at Highfield, Court Lodge, Bayshaw, or elsewhere in the valley drifts near Ightham. At Aylesford he found one, and at Milton Street none.

A few derived specimens have been found at Seal, Chart Farm, Stonepits, and Milton Street.[2] Mr. H. Lewis has also sent me one of the double-curved scrapers (the depressed form), characteristic of the plateau group, which he found in the Limpsfield Common gravel pit; and Mr. Bell informs me that he has certainly seen and rejected some of the older rude specimens, owing to the absence of the bulb of percussion which he then considered essential. That a

---

[1] I have one specimen, four inches long, given me by M. Boucher de Perthes, from near St. Riquier, five miles north east of Abbeville, which may belong to this group. It is said to have been found at a depth of four mètres, and apparently comes from a red clay drift, which there lies on the top of the higher chalk hills.

[2] Mr. Lewis has, however, found a rather larger proportion of them in the Milton Street pits, and one still larger in the gravel at Southfleet, and Mr. Bullen has several specimens from the pit at Aylesford.

certain number of derived specimens should be found in the valley gravels is to be expected, if we bear in mind the extent of surface lost in the erosion of the plateau by the valleys which traverse and flank it on all sides. While the chalk in these valleys was washed away and lost, the harder fragments of flint and chert have been left behind amongst the general *débris*, or in the reconstructed valley gravels. The derived plateau specimens are easily distinguished, by their greater wear (on the whole), distinct colour, and peculiar shapes, from the implements contemporary with these valley drifts.

I do not wish to assert that all the plateau implements are of so distinct a pattern that they can always be distinguished from the valley implements. Nor would I insist that their rudeness alone is a proof of their antiquity, for not only are there rude specimens in the valley drifts, but the plateau group includes a certain number of prototypes of the common valley types. The difference consists in the fact that the great majority of the plateau specimens are of rude make and of peculiar types, while well or even moderately well-finished specimens are rare, and also in that they possess special physical and structural characters, in accordance with a presumed greater geological antiquity. The presence of a few forms in common does not invalidate the general evidence any more than does the circumstance of the implements in low-level valley drifts being scarcely distinguishable, except perhaps in the proportion of some forms, from those of the high-level gravels, notwithstanding the difference of position and age.

4. Distinctive Characters of the Valley and Plateau Implements

But the special question which I wish to bring before you is whether, taken as a whole, the plateau implements exhibit distinct characters and types, such as would denote them to be the work of a more primitive and ruder race than those fabricated by Palæolithic Man of the valley-drift times. The chief features of the valley implements may briefly be stated as follows :—

The Valley Implements.—With respect to these, it is manifest that the great majority of the implements have been made from large flints, either taken direct from the chalk or else found lying on the surface or in the bed of streams. I have before cited the interesting case at Crayford, described by Mr. F. C. J. Spurrell,[1] where at the foot of a submerged Chalk cliff on the old Thames bank he found, beneath a depth of twenty to thirty feet of Mammiliferous drift beds, the surface of a former strand strewed over with the flakes and chips of large flints, of which a layer is still exposed in the adjacent old cliff. The fragments are as sharp and fresh as those of a recent Norfolk flint-knapper's shop, and show as well-marked bulbs of percussion, and though they are scattered about in disorder, they often admit of being put together again in a manner to restore the form and substance of the original blocks. Other old workshops

[1] *Quart. Journ. Geol. Soc.*, vol. xxxvi. p. 544.

away from the chalk area, and therefore dependent upon flints from the surface, or from the bed of streams, have been found by Mr. Worthington Smith[1] at Stoke Newington, and by Mr. Allen Brown[2] at Acton.

A few only of the valley implements are made from gravel-flints, or from the green-coated flints at the base of the Lower Tertiaries. The size of the implements varies generally from 3 to 6 or 7 inches in length, but some specimens attain a length of 10 to 12 inches. They are in general but little worn, though there are exceptions to this rule. The workmanship of the implements often shows considerable skill, and some of the smaller lanceolate forms are chipped and finished with a neatness approaching to the Danish javelins or spear-heads of the Stone period. At the same time there is a considerable proportion present of rude and sometimes very rude specimens, but they all belong to the same types as the more finished specimens. Some of the implements may have been used in the hand, but a large and probably the greater number must have been fixed or tied to handles of wood or bone.

Sometimes the colour of the flint of which the implements are made is in no degree altered, but remains as black and fresh as originally. At other times it is stained yellow or ochreous by the matrix in which it has been embedded, or else, owing to a molecular change of the surface, it has assumed the white colour and glazed aspect (*patina*) of porcelain.

[1] *Journ. Anthrop. Inst.*, vol. xiii. p. 357.
[2] *Palæolithic Man in North-west Middlesex*, pp. 56–60.

The following are the principal forms of the valley implements :—

1. *Narrow flakes with sharp untrimmed edges*—for cutting.
2. *Oval, leaf-shaped, and round flakes,* worked on one side or at the edges—scrapers.
3. *Pointed implements* with the haft end formed by the natural surface of the flint—tools or weapons.
4. *Pointed or lance-head implements,* often very large; and worked entirely out of the flint—same as 3.
5. *Spatula-shaped implements,* generally all worked out of the flint, and often with both ends sharp and cutting,—adzes, axes, hatchets.
6. *Flat ovoid implements* worked all round.
7. *Ovoid implements* with a slight twist in centre.
8. *Flints used as hammer stones.*

Of these the pointed forms, Nos. 3, 4, and 5, largely predominate, and next are the ovoid forms, Nos. 6 and 7.

The Plateau Implements.—The characters which on the other hand distinguish these implements are :—

1. They are almost all more or less stained, like the broken drift-flints with which they are associated, of a deep warm brown colour, which spreads alike over the natural and the worked surfaces, although in some cases it is lighter on the latter. Some have specks of iron oxide attached.
2. They also, like the natural flints, generally show a considerable amount of wear, as though they had been a good deal rolled and knocked about, so that the worked edges are commonly rounded off and blunted. This character combined with the next often renders the artificial work very obscure, especially as it is very rarely that a specimen shows the bulb of percussion.
3. The trimming is also often very slight, and has generally been made on the natural edges of broken flints which were picked up on the surface. It is only in a very

few instances that an implement has been entirely wrought out of a larger flint, whereas, as just mentioned, with the valley implements such is commonly the case. The trimming, slight though it may be, is to be recognised by its being at angles or in places incompatible with river-drift agencies, and such as could not have been produced by natural causes, which tend to remove all projecting points.

4. Besides the implements of definite patterns, there is a large, probably the larger, number, which though not the result of chance, show no special design. Amongst these are the natural flints which have been selected for use as hammer or trimming stones, the result being that the flint has become chipped at the ends or round the sides, as it were undesignedly, but still in a manner that could not have resulted from natural wear. In a similar way, some are roughened at the end like the large round pebbles used at a later Neolithic period, exhibiting patches of rough abraded surface, the result of repeated blows.

Attention to these points will soon enable one to distinguish between a natural flint, however they may otherwise simulate the artificial forms, and those flints which show an object to be attained, in however simple a way that object has been carried out.

But although the great majority (at least 95 per cent.) of the specimens are of rude primitive forms, there are some which might pass as indifferent valley-drift specimens of the ovoid and pointed types, while a few large implements have been found equalling in workmanship and finish some of the best of the valley specimens. One specimen (mentioned at p. 56) from South Ash, in Mr. Harrison's possession, is most carefully fashioned. It is 6 inches long by $3\frac{1}{2}$ inches in width, and is of the thin flat spatula-shaped form, and of a bright yellow colour. Two others are large flat ovoids from a field off Bower Lane, whilst another, also from Bower Lane, has the top broken off, and is similar in shape to the Ash specimen, but is weathered white, as though it had not been entombed in drift.

Two well-made implements of the sharp-pointed St. Acheul (Amiens) type have also been found— one, white, is from Kingsdown, and the other, stained yellow, is from Ash. Neither of them are at all worn, and both show a slight *patina*.

It is not easy to account for the presence of these abnormal specimens. If contemporaneous with the others, we might assume that there were then some workmen more skilled than their neighbours in the fabrication of flint implements. But if so, how is it that they are not more numerous, and that there should be so great a difference between those and the other forms, or that there should be few or no intermediate forms? How also are we to account for the great difference in physical conditions? The rude specimens which preponderate so largely are of a nearly uniform brown colour, and are usually much worn; while the few rare finished specimens are sharp, show no wear, and are of lighter tints.[1]

Again, we may suppose that the plateau heights were at the later Palæolithic periods frequented by man of the valley drifts for the purpose of the chase or to supply some want, and that some of their implements were lost when in pursuit of game, or in felling trees, or grubbing up roots. But these are

[1] It seems most probable that they belong to different times.

mere conjectures, and we must wait in the hope that some new sections will throw further light upon the geological question, and show with greater certainty the exact stratigraphical relation of these different types of implements.

Whilst also the more finished implements, which are identical in make with those of the valley, are unworn, and have all the appearance of having remained *in situ*, the rude implements are worn, and would appear to have been carried down, with a foreign drift, on to the plateau, from those Central Wealden uplands which I have estimated might, in Preglacial times before the denudation of the Weald, have formed a low mountain range 2,000 to 3,000 feet in height. The reason for supposing that these rude implements have been thus swept down from those uplands is that on these chalk hills worn fragments of chert and ragstone, *derived* from the Lower Greensand hills to the south, are widely scattered, and the implements are associated with them. At the same time there is nothing to indicate that they were made on the spot where they are now found. It is possible even that they may have to be relegated to a still earlier period than I have named. For the present I am concerned only to show that their extreme rudeness, as a whole, points to a very primitive state of art, such as would be in accordance with the antiquity indicated by the geological evidence.

[1] *Quart. Journ. Geol. Soc.*, vol. xlvi. p. 169.

5. Typical Forms of the Plateau Implements

Unlike the Valley Implements, the Plateau Implements are, as a rule, made of the natural *fragments* of chalk-flints, that were found scattered over the surface of the ground, or picked up in beds of gravel, and merely roughly trimmed. Sometimes the work is so slight as to be scarcely apparent; at others, it is sufficient to show a distinct design and object. It indicates the very infancy of the art, and probably the earliest efforts of man to fabricate his tools and weapons from other substances than wood or bone. That there was an object and design is manifest from the fact that they admit of being grouped according to certain patterns. These are very simple, but they answered to the wants of a primitive people.

With few exceptions the implements are small, from two to five inches in length, and mostly such as could have been used in the hand and in the hand only. There is, with the exceptions before named, an almost entire absence of the large massive spear-head forms of the Valley drifts, and a large preponderance of forms adapted for chipping, hammering, and scraping. With these are some implements that might not have been used in the hand, but they are few and rude. The difference between the Plateau and the Valley implements is as great or greater than between the latter and the Neolithic implements. Though the work on the plateau implements is often so slight as scarcely to be recognisable, even modern savage work, such as exhibited for example by the stone implements of the Australian natives, show, when divested of their mounting, an amount of work no greater or more distinct than do these early palæolithic specimens.

Some persons may be disposed to look upon the slight and rude work which these flints have received, as the result only of the abrasion and knocking about caused by collision during the transport of the drift. This belief prevailed for a time even in the case of the comparatively well-fashioned valley implements. A little practice, and comparison with natural drift-flints, will show the difference, notwithstanding the, at first, unpromising appearance of these early specimens of man's handicraft. It is this fact of their being the earliest such work with which we are at present acquainted that lends them so great an interest, for they give us some slight insight into the occupation and surroundings of the race by whom they were used. A main object their owners would seem to have had in view was the trimming of flints to supply them with implements adapted to the breaking of bones (for the sake of the marrow); or for scraping skins and round bodies—such as bones or sticks—for use as simple tools and poles. From the scarcity of the large massive implements of the pointed and adze type, so common in the Valley drifts, it would seem as though offensive and defensive weapons of this class had not been so much needed, whether from the rarity of the large Mammalia so common later on in the low-level Valley drifts or from the habits and character of those early people. It would be interesting to institute a more careful comparison of these rude implements with the stone implements of different modern savages, to ascertain what light it could throw upon their use, or whether they point to entirely different habits and usages.

In order to form some estimate of the proportional numbers and character of the type specimens, I have carefully collated the large collections of Messrs. Harrison and De B. Crawshay, which they kindly placed at my disposal for that purpose, and for comparison with the implements from the Valley drifts in my own and other collections. The following grouping embraces the main forms typical of this Plateau drift. Some of the forms pass, however, so insensibly into others, that it is difficult to draw the line between them, or to say whether more divisions should not be made.

The greater number of the implements, no doubt, are so rude that probably few of us would agree upon a general classification. On the other hand some types are peculiar and very distinct, while a few are common to the later valley drifts, such as the pointed St. Acheul and the ovoid Abbeville types; but with few exceptions all the others are of more simple types, and have a stamp of their own, and most of them seem confined to this epoch. They may, for general purposes, be divided into three groups: (i.) The first consisting of those in which the natural flint has been used with little modification, and in which the original shape of the stone has determined the ultimate form. (ii.) In the second the natural flints were equally used, but design is more apparent in adapting them to a common pattern and special purpose. (iii.) In the third, the implement or tool has been worked more or less entirely out of the flint with a definite object in view, as is the case with the later palæolithic implements.

The first group (Pl. I., II.), which is by far the most numerous, includes the following types:—

1. *Thin natural flat pieces of flints*, with merely the sides chipped and notched, sometimes all round, and sometimes brought to a rude point. This includes a variety of forms but of no definite pattern (Figs. 1, 2, 3).
2. *Split Tertiary flint-pebbles*;[1] chipped round the edges for cutting or scraping (Fig. 4).
3. *Larger flints that fit the hand* and could be used with a little dressing as trimming and hammer stones. These show the notches, the result of blows given by use, and may have served to shape other flints, or to break bones or other hard substances. They are extremely numerous and variable (Fig. 5).

The second group (Pl. III. to IX.), which might almost be included as a sub-group in the first group, is also very numerous, and includes a greater variety of more definite forms, especially scrapers of various patterns, such as—

4. *Ordinary Scrapers*, formed out of natural flakes or flat pieces of flint trimmed at the edges (Figs. 6, 7).

[1] I learn from Dr. H. P. Blackmore, of Salisbury, that similarly made scrapers were still in use with some of the American Indians, who, whenever in want of a scraper, select a pebble, which they split, and then trim off the edges. They rarely keep the old scraper, fresh ones being so easily obtained. I annex a reduced copy of his sketch. It is called a *pashou*, or scraper, and is used by the Shoshone Indians to dress skins. Obtained by Professor Joseph Leidy, at Fort Bridger, Wyoming, in 1872. A similar plateau implement is figured on Pl. II., Fig. 4.

5. *Knob-headed and shoe-shaped Scrapers,* formed from a natural rough flint, flat on one side, and trimmed at one end (Figs. 8, 9).
6. *Massive thick broad Scrapers,* flat on one side and trimmed round one or two edges.
7. *Square-headed or chisel-shaped Scrapers.*
8. *Crescent-shaped Scrapers.* These are generally small and adapted to scrape a round body such as a bone or a stick.[1] This is a very characteristic form (Figs. 11, 12, 13, 14, 15).
9. *Double Scrapers.* These seem intended for the same object, but have two scraping edges with an intervening point. This is another characteristic form (Figs. 16, 17, 18, 19).
10. *Bow-shaped Scrapers with a small central point* (Figs. 20, 21, 23, 24, 25). Very characteristic. Fig. 22 is made from a flat pebble.
11. *Double Scrapers in the form of an hour-glass.* These are generally formed out of a thin flat natural flint, and are not common.
12. *Drills, Borers, Wedges?* (Figs. 26, 27, 28, 29, 30).
13. *Beak-shaped Implements* which may possibly have been used as scrapers or as picks. These also are characteristic (Figs. 32, 33, 34, 35).

The third group (Pl. X., XI.) includes forms common in the Valley drifts but comparatively rare among the Plateau implements, and comprises—

14. *Plain Flakes, occasionally showing the bulb of percussion,* used for cutting purposes, and so common in the valley drifts, but very rare here.
15. *Broad Flakes,* trimmed round the edges (Fig. 36).
16. *Ovoid Implements* of the same pattern as those common at Abbeville (Fig. 39).
17. *Pointed Implements* of the Spear and Lance-head pattern (a St. Acheul type). These are small and rough compared with those in the valley gravels (Fig. 38).
18. The same with a slight curve at point (Fig. 37).

---

[1] These resemble in shape an instrument termed a "draw shave," used in Kent for shaving hop-poles (Harrison).

Besides these there are other forms of which only single or few specimens have been found. Amongst them are a few rounded flints, which may have been used as Fling-stones; others are crook-shaped.[1]

But although a certain number of the specimens admit of classification, a very large number might pass under different headings, and have evidently been made to serve more than one purpose. The most common combination is that which combines any one of the described types with the several forms of scraper. Examples of these may be seen in Figures Nos. 7, 20, 22, 23, 24, 25, 31, 33, 34, 35, and others. Though the material was so common, these early men seem to have economised it as far as possible by trimming any suitable flints into tools alike for hammering, or chipping, or scraping skins, sticks, or bones, or for boring or drilling purposes, so as to combine in one tool two or more objects. This is a feature peculiar to the Plateau implements.

The relative proportion in which these groups occur may be approximately estimated at—

Slightly worked flints of the 1st group, 50 per cent.
Designed implements ,, 2nd ,, 44 ,,
Implements of more finish, 3rd ,, 6 ,,

It is probable, however, that the first group is under-estimated, as so many of the specimens were not considered worth keeping, and were thrown away.

[1] Mr. Harrison has adopted a somewhat different grouping 1. *Crook-point Tool*; 2. *Single-curve Scraper*; 3. *Double-curve Scraper*; 4. *Combination Tool*; 5. *Split-pebble Tool*; 6. *Semi-circular Tool*; 7. *Drawshave* or *hollow Scraper*; 8. *Tool with work all round*. See *Journ. Anthrop. Inst.* for 1892, p. 265.

The greater number of the second group consist of the various forms of scrapers.

No estimate has been made of the proportions in which the different types of the later palæolithic implements occur in the Valley drifts; but probably not less than one half consists of the different forms of the Pointed and Ovoid types, and one quarter of the different forms of flakes.

Looking at the very distinctive features of the Plateau Implements, such as their rudeness of make, choice of material, depth of wear and staining, peculiarity of form—taken in conjunction with the extreme rarity of the valley forms—these constitute characters so essentially different from those which typify the Valley implements, that by those characters alone they might be attributed to a more primitive race of men. As this view also accords with the geological evidence which shows that the Drift beds on the Chalk plateau, with which the implements are associated, are older than the valley drifts, I do not see how we are to avoid the conclusion, that not only was the plateau Race not contemporary with the valley Men, but also that the former belonged to a period considerably anterior to the latter—possibly an early Glacial or even a Preglacial period.

Since the date of the reading of this communication, Dr. E. Tylor's interesting paper "On the Tasmanians as *Representatives* of Palæolithic Man" has been published.[1] It would appear that the Tasmanians up to the time of the British colonisation

[1] *Journ. Anthrop. Inst.*, Nov. 1893, p. 141.

"habitually used stone implements shaped and edged by chipping, not ground or polished." Dr. Tylor states that they are of the same character as the palæolithic implements of "the Drift and Cave periods," especially like those characteristic of the cave of Le Moustier. There are, likewise, amongst the specimens he figures several which will compare with the plateau implements of Kent. Figs. 2, 3, 5, 6, Plate X., might be matched with such forms as those which I include in type 2, Fig. 3, whilst Fig. 1, Plate X., will compare in general character with my Fig. 4, Plate VIII., in the *Quart. Journ. Geol. Soc.*, vol. xlvii.

On the other hand no implements resembling the common Spear or Lance-head type of later palæolithic age, and so common in the Somme and Thames valleys, have been found in Tasmania. The modern savage, like the ancient one, "is content to use a few forms of implements for all purposes of cutting, chipping, &c., those being flakes as struck off the stone, and such flakes or even *chance fragments* trimmed and brought to a cutting edge by striking off chips along the edge of one surface only, whether completely or partly round." "There is no record of the small chips ever being flaked off by pressure but only *by blows with a stone*," and this appears to have been the case with the plateau implements.

Dr. Tylor also states that according to "experienced observers the implements were *grasped in the hand* for use, never mounted on handles." This again agrees with the conclusion we had formed respecting the plateau implements.

In the discussions to which this subject has lately given rise, it has been contended—

1st, That there is no sufficient evidence of the age of the Plateau drift, or that it is older than that of the valleys.

2nd, That the so-called implements are natural and not worked forms of flint.

With regard to the first point, as no definition of the objections has been risked, I can only refer to the section at p. 53, *ante*, and to the more detailed sections I gave in a former paper.[1] These give the evidence on which my opinion of their age is founded, and which has not been refuted. Not only is there a great difference of level, but, whereas the River-valley drift is derived entirely from strata occurring within the river basins, the Plateau drift is derived from strata beyond the circumjacent valleys, at a time when these valleys did not exist. It is clear that the *Chert* and *Ragstone* present in the Plateau drift travelled from the Lower Greensand of the hills to the south, when the intervening valley was bridged over by strata since denuded. Though closely associated with the "red clay with flints," the Plateau drift is in no way connected with it except by superposition or infraposition.[2] It however overlies in places a bed of Pliocene (Lenham) age.

[1] *Quart. Journ. Geol. Soc.*, vol. 47, p. 125, and Pl. VI. Fig. 2.

[2] A grant has recently been made by the British Association to enable Mr. Harrison to have some pits dug in order to determine more exactly the relation of those beds and the depth to which the Plateau implements are found. With respect to the last point, it is satisfactory to find that they have been met with at a depth of five to six feet. Other pits will be sunk (Nov. 1894).

The independence of these two deposits has lately been confirmed by some observations of Dr. Blackmore in the neighbourhood of Salisbury. Alderbury Hill, an isolated hill of London clay, free from the "red clay with flints," is capped by an ochreous gravel, in which, at a depth of 11 feet, Dr. Blackmore has found several well-marked implements of the rude Plateau types. The Valley gravels are largely developed in the adjacent valley of the Avon, at a level of about 300 feet lower.

On the second point, there may have been more reason for hesitation. The majority of the specimens are so extremely rude and the variety of forms so great that until a particular form is found to be constantly repeated, the evidence of work might easily escape recognition.[1] Rude, however, as the specimens are, there are touches in the trimming which it is not possible to refer to natural causes, but have evidently been made by one having an object in view. Rudeness of form by itself is, however, no test of age. There are rude implements of the Valley gravels, as well as of Neolithic times, whilst among the stone implements of living savages there are many as rude as those of the Plateau group.

But each epoch had its typical forms, and these are broadly persistent, however rude the specimens may be. In the Neolithic period, axe and chisel shapes predominate; in the Valley gravels the long pointed and spatula-shaped implements are characteristic of the period; and in the Plateau group, various forms for

[1] A former generation of geologists was for a time equally incredulous respecting the valley specimens.

scraping and hammering prevail. There are no doubt pointed forms in the Plateau group—prototypes of the later implements—but they may have a different *cachet* from those of the Valley group, as those again have from those of the Stone period. There are also certain generalised forms which persist throughout all the periods, though perhaps varying a little in some minor details. Simple flakes likewise, more or less worked, are common to all three periods. It is true that some specimens found on the Plateau are as well worked as any in the Valley group, and how to account for their presence yet presents some difficulty, but that they are not of the same age, I feel nearly certain. Not only is their make different, but their condition; their freshness—if it may be so termed,—and their rarity constitute differences so great, that placed side by side, they would never be placed in the same category. That they should be found on the plateau is no more surprising than that unmistakable Neolithic implements are found in the same surface in company with the Plateau implements. The characters of the latter two are however so distinct that no one could question their relative age; and so it is with the older two sets.

A more untenable objection has been raised that the flints cannot have been worked, because they show no *bulb of percussion*. Did it never occur to the objectors that hundreds of the Valley specimens show no bulb. When the trimming is produced by slight chipping or by pressure there can be none. The bulb of percussion is however to be found on Plateau specimens, though more rarely in the Valley drifts.

Though a certificate of handiwork, its absence is no proof to the contrary.

It is in <u>the frequent repetition of the same form,</u> as seen in a series of them, that the evidence of an intelligent design strikes one most forcibly and convincingly. That hundreds of flints should have assumed the same general form without the intervention of human agency, is so highly improbable that it approaches to an impossibility. Can those who assert that they are chance natural forms prove it by showing natural flints having the forms of the Plateau specimens? Though challenged to do so, not one such specimen has been forthcoming in the three years since this paper was read. If also the flints are supposed to be formed by natural causes, they ought to be found in all drift gravels, and not confined to this particular bed. The tendency, moreover, of all river and sea wear is to rub off all angles, and reduce the stone to a round pebble.

I think that few persons could look through the series of the more typical forms in Plates I. to XII. without feeling that they have been fashioned with evident intent. Although there are hundreds of specimens having very indefinite forms, a large majority of these will still be found to have relations more or less distinct (often very faint) with these types. It is evident to me that we are here in presence of a very simple and may be nascent intelligence. The work is in fact such as we might expect from a race of a time so remote from us and so remote even from Palæolithic Man; for whereas at the time of the Valley-gravels this land had assumed its present main physiographical features of hill and dale, at the time of the Plateau drift the chief of these surface features were non-existent. We may judge from this how great was the distance of time which separated the two races.

Decided opinions, adverse to Plateau Implements showing work,[1] have been expressed, but in greater part by geologists who are unacquainted with the ground and little experienced in this branch of research. I am satisfied however that if they would spend one day on the Plateau they would not fail to see the difference between the natural and artificial forms of the flints scattered over the surface. Even the labourers who aid Mr. Harrison in his search rarely make a mistake. But the eye requires a certain education which is not to be acquired by discussion or in the cabinet.

On the other hand, archæologists the best able to form an opinion on this subject, such as Dr. Blackmore and Canon Greenwell, maintain, with us, the authenticity of the specimens. I have the permission of Canon Greenwell to record his belief " in the manufacture by some reasoning creatures of the flints in question "; .... that he " has not the least doubt of their having been made with intent "; .... and that he " knows of no natural agency which has or indeed could produce the signs of work so abundantly shown upon them." No other conclusion is, to my mind, possible.

[1] The finish and symmetry of the finer Valley and Neolithic specimens would seem to have established a standard of art under which the rude and uncouth Plateau specimens find no place. A different standard is needed.

PRIMITIVE CHARACTERS OF FLINT IMPLEMENTS 79

EXPLANATION OF PLATES.

Plate I. *First Group.*

Fig.
1 (B) Cockerhurst, Shoreham ....⎫
2 (190) West Yoke .... ⎬ Type No. 1
3 (51) Fairseat ....⎭

Plate II. *First Group.*

4 (1,688) North Ash..................................... Type No. 2

23 (2,630) West Yoke ....⎫
24 (1,178) Turner's Oak..............................⎬ Type No. 11
25 (3,313) West Yoke ....⎭

Plate VIII.

26 (2,536) Ash	⎫	
27 (B) Sepham Heath, Halstead	⎬ Type No. 13	
28 (316) South Ash	⎭	
29 (316) South Ash.............................	⎫ Type No. 14	
30 (520) Ash	⎭	

Plate IX.

31 (3,317) West Yoke .... Type No. 15
32 (2,159) Ash .... ⎫
33 (1,615) Ash .... ⎬
34 (77) West Yoke.. .... ⎬ Type No. 16
35 (3,439) West Yoke .... ⎭

Plate X. *Third Group.*

36 (375) Ash .... Type No. 17
37 (561) Ash .... Type No. 18

Plate XI.

38 (661) Barnshatch .... Type No. 20
39 (173) West Yoke .... Type No. 19

Plate XII.

40 (2,117) North Ash............................. Type No. 11
41 (1,384) Barnshatch .... Type No. 9

The numbers in brackets are those of specimens in Mr. Harrison's collection, except those with B prefixed, which are in the collection of the Rev. R. Ashington Bullen.

A map of the district showing the position of the several places mentioned in the above list will be found in the *Journal of the Anthrop. Inst.*, Vol. 21, Pl. 18.

The specimens are all, with few exceptions, drawn to natural size.

Pl. XII. shows two unusually large specimens of the Second Group.

The artificial character of the chipping on the edges and points of these specimens is well shown on most of the figures by their greater sharpness and localisation, while the natural surfaces remain rough and with edges unchanged.

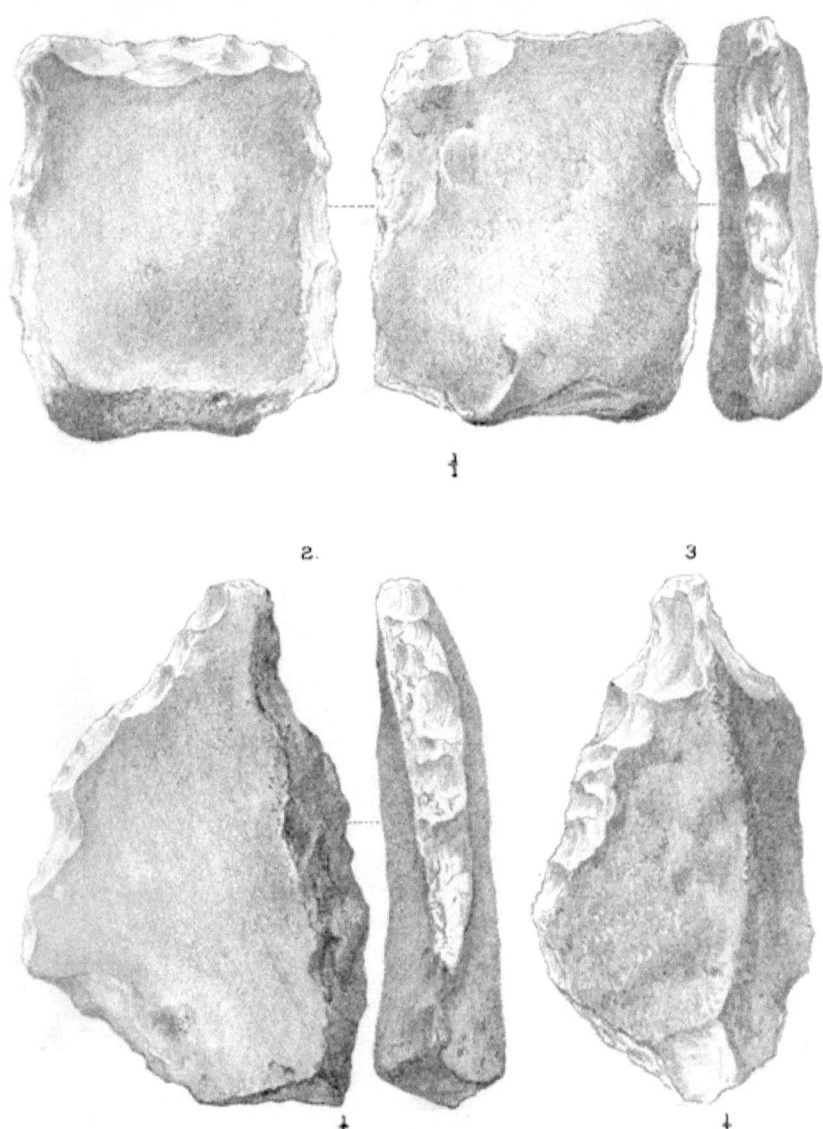

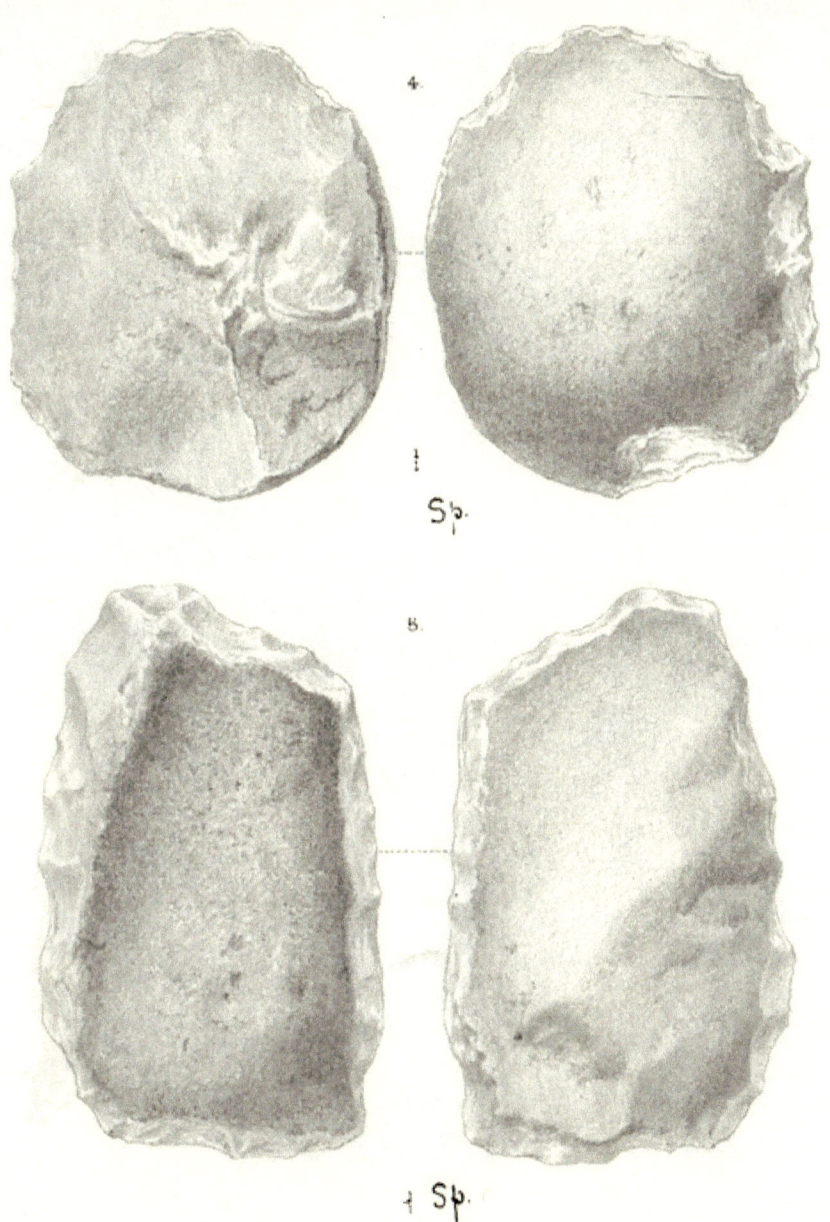

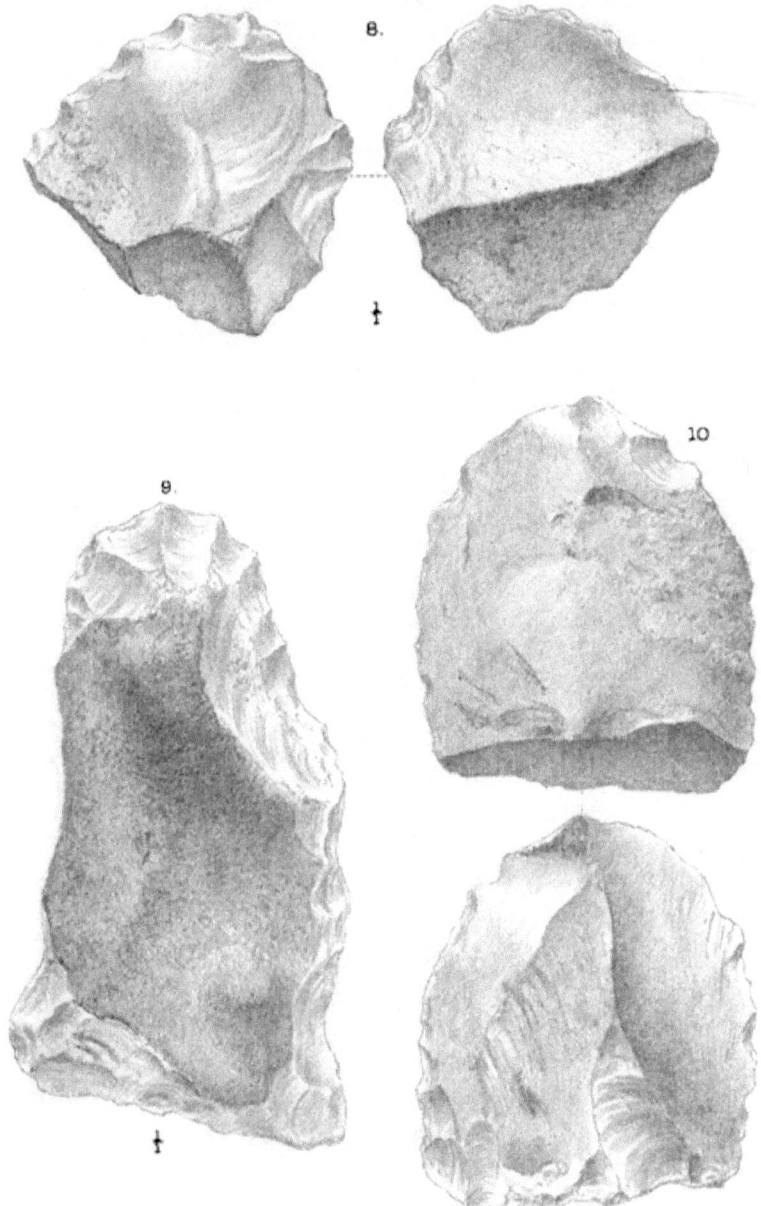

SECOND GROUP. Plate V.

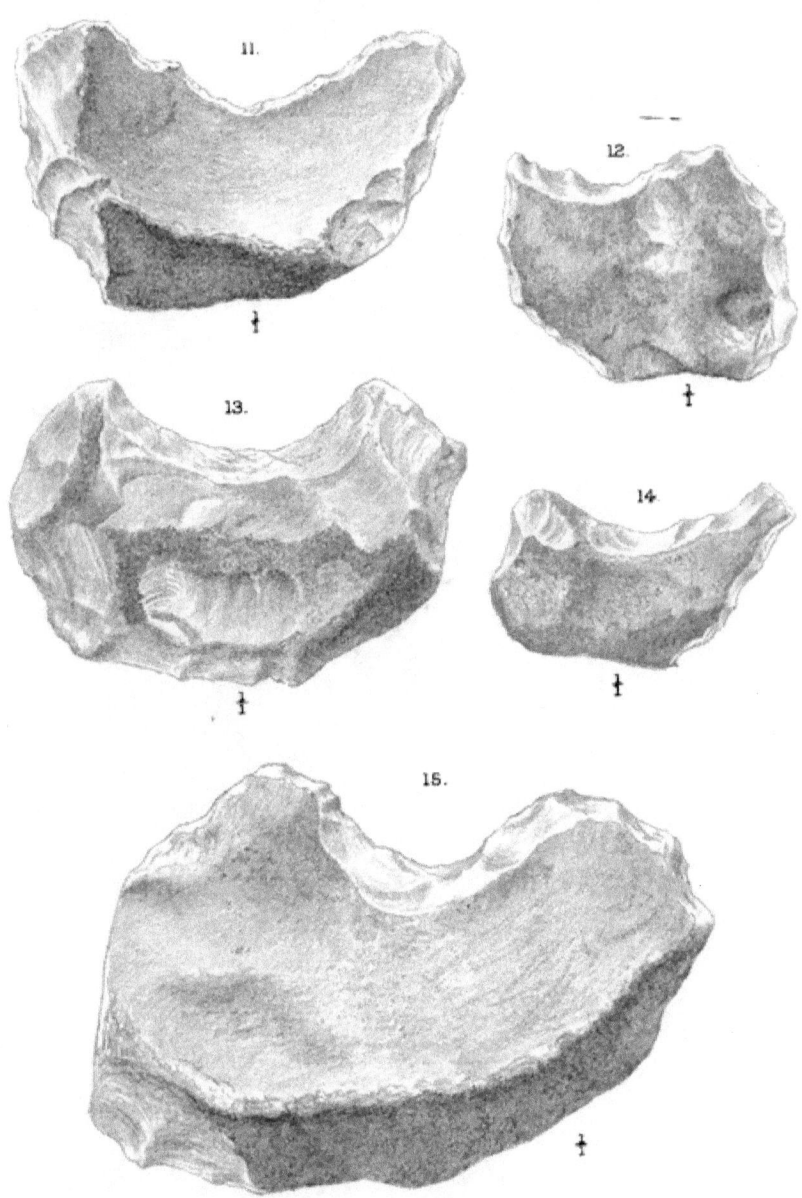

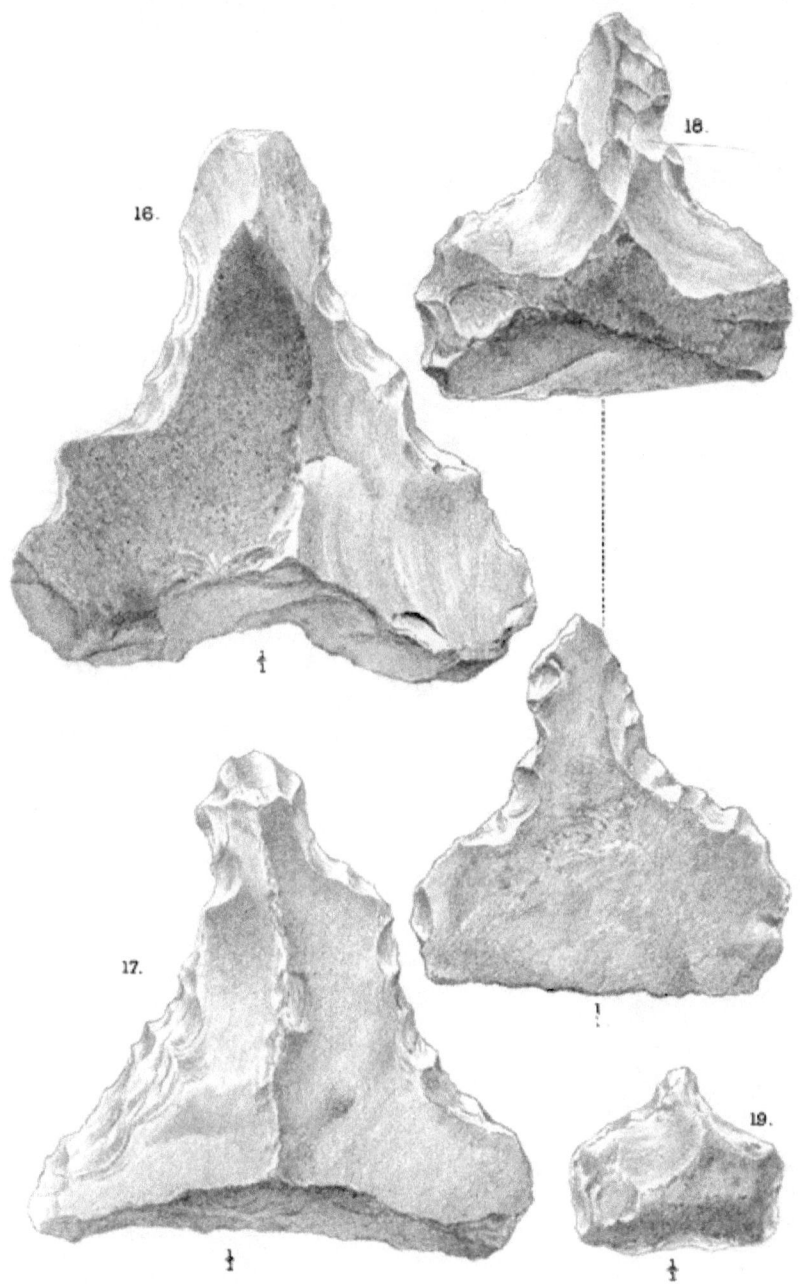

Plate VII.

SECOND GROUP.

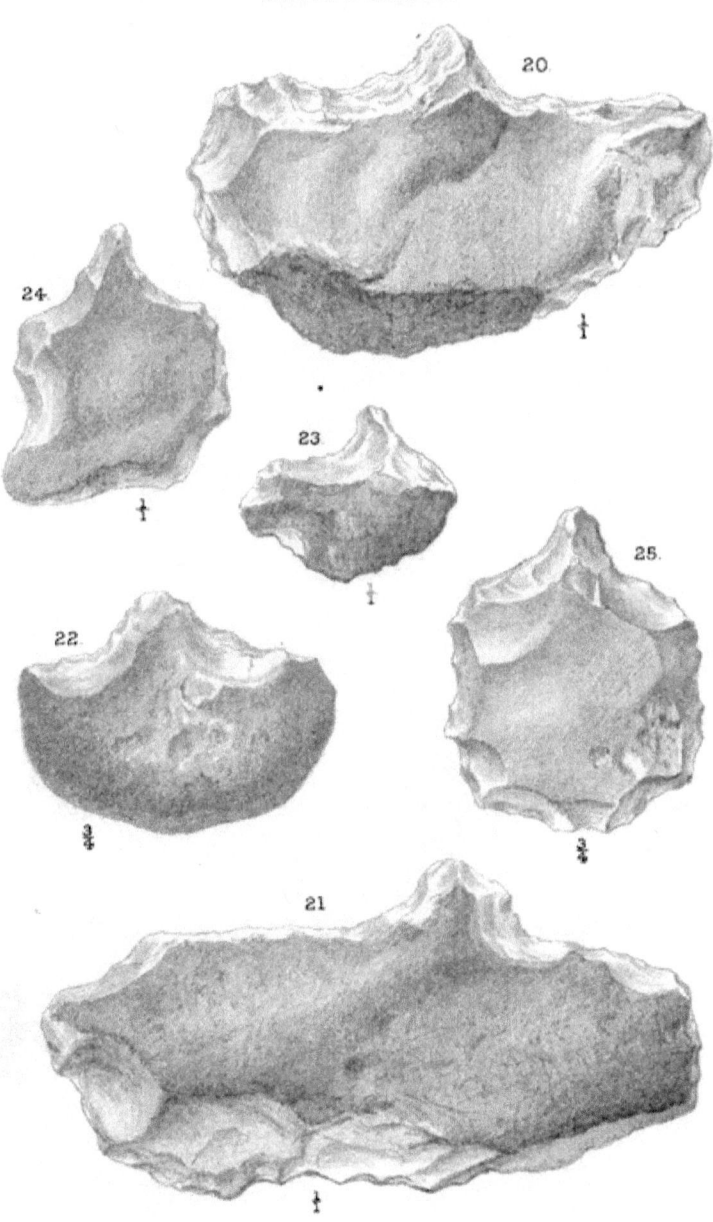

SECOND GROUP. Plate VIII.

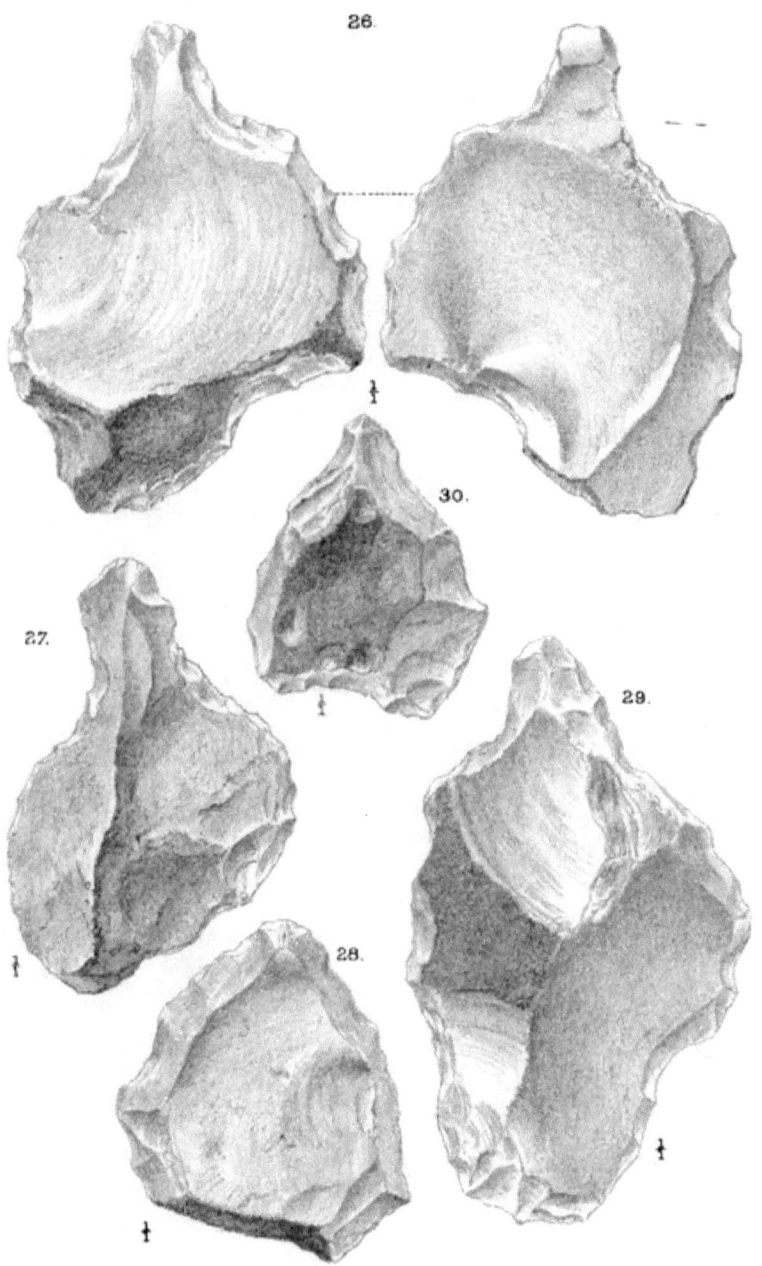

SECOND GROUP. Plate IX.

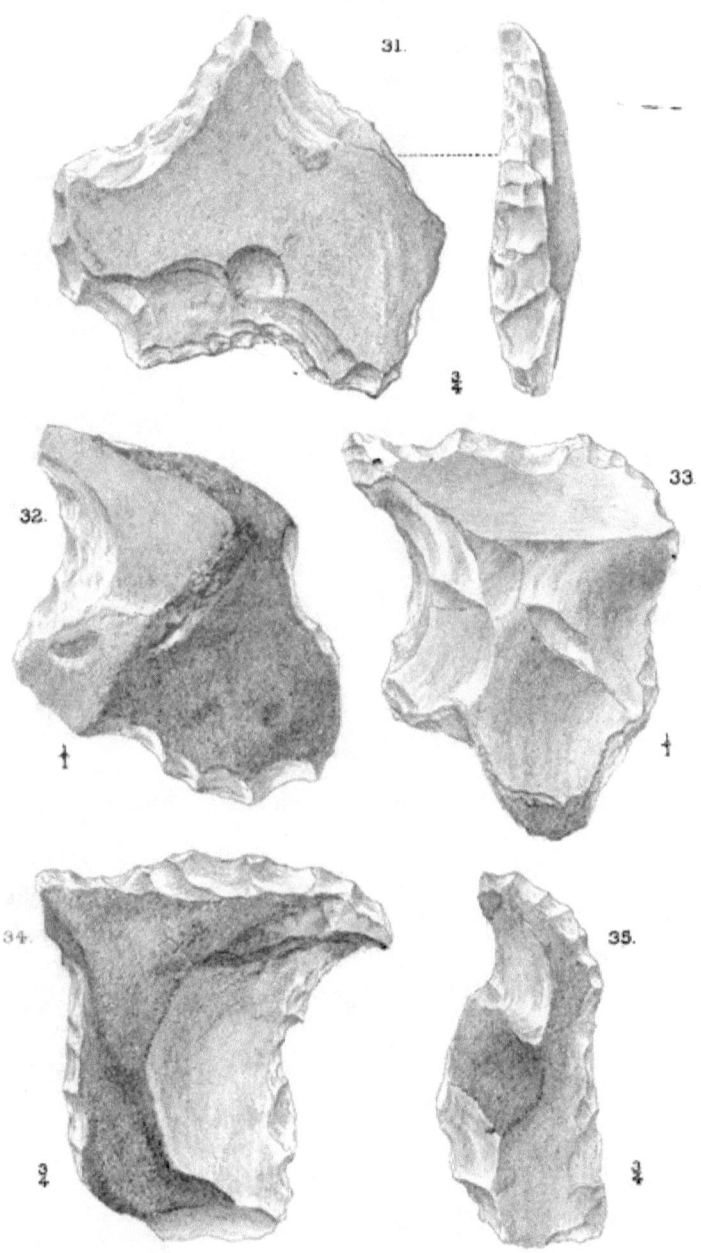

Plate X.

THIRD GROUP.

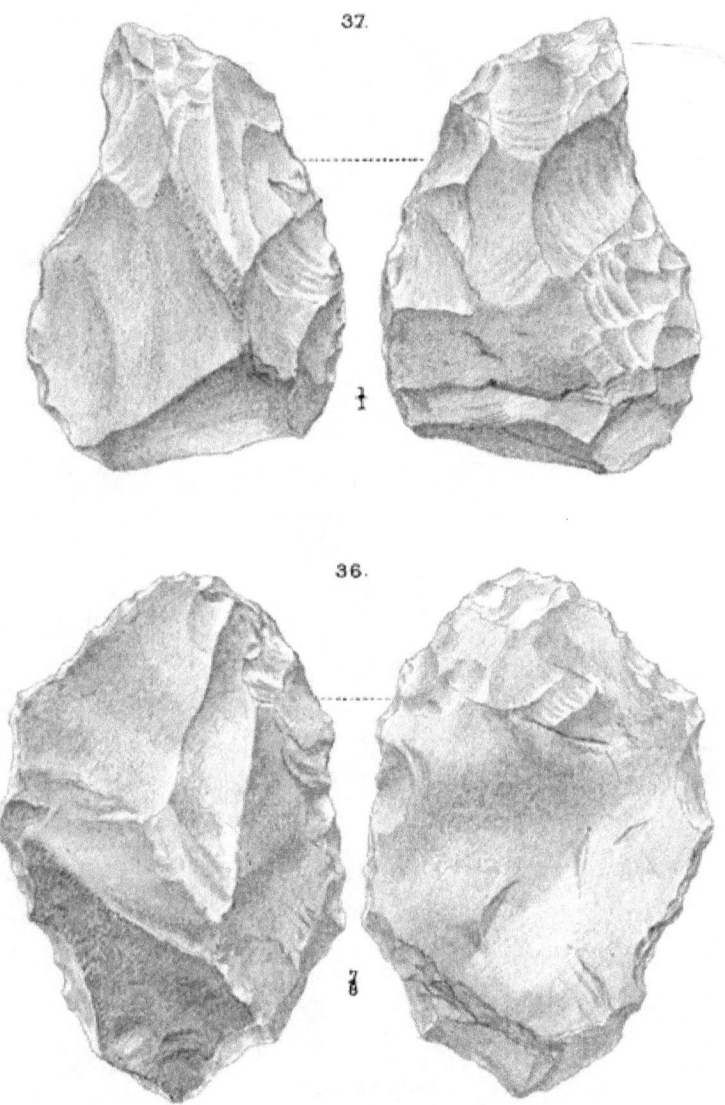

THIRD GROUP. Plate XI.

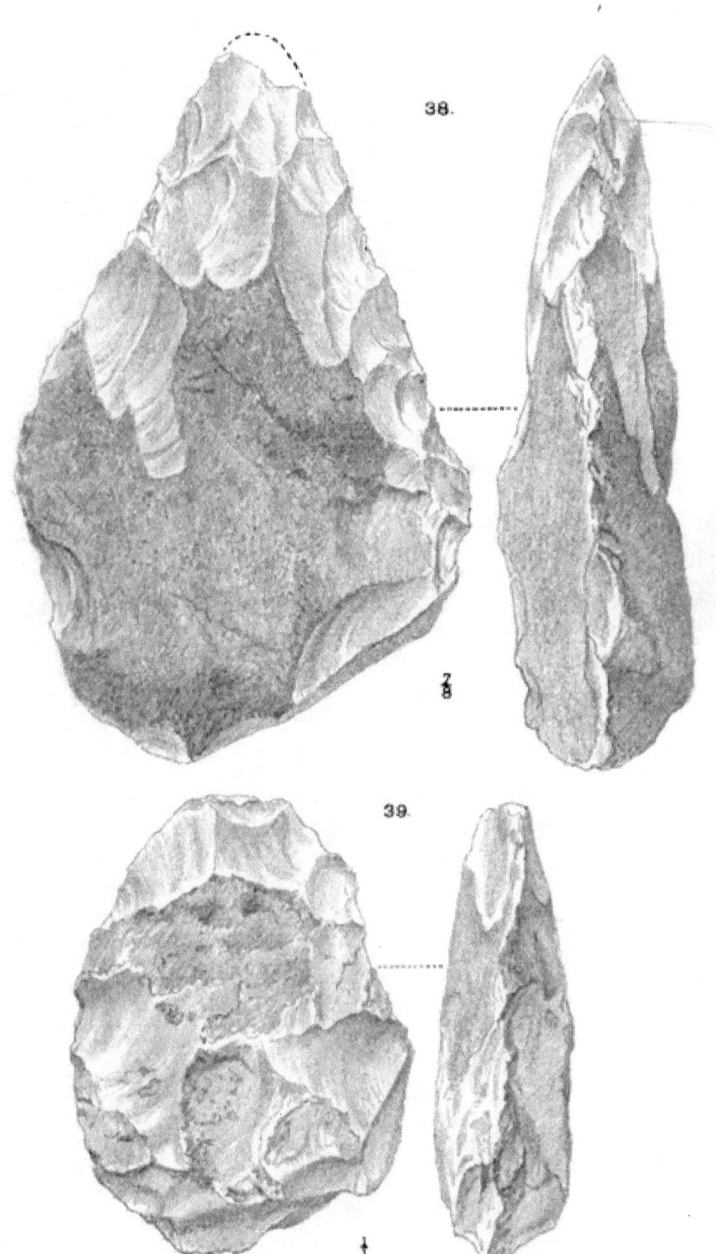

W S Tomkin del & lith.

SECOND GROUP.

Plate XII.

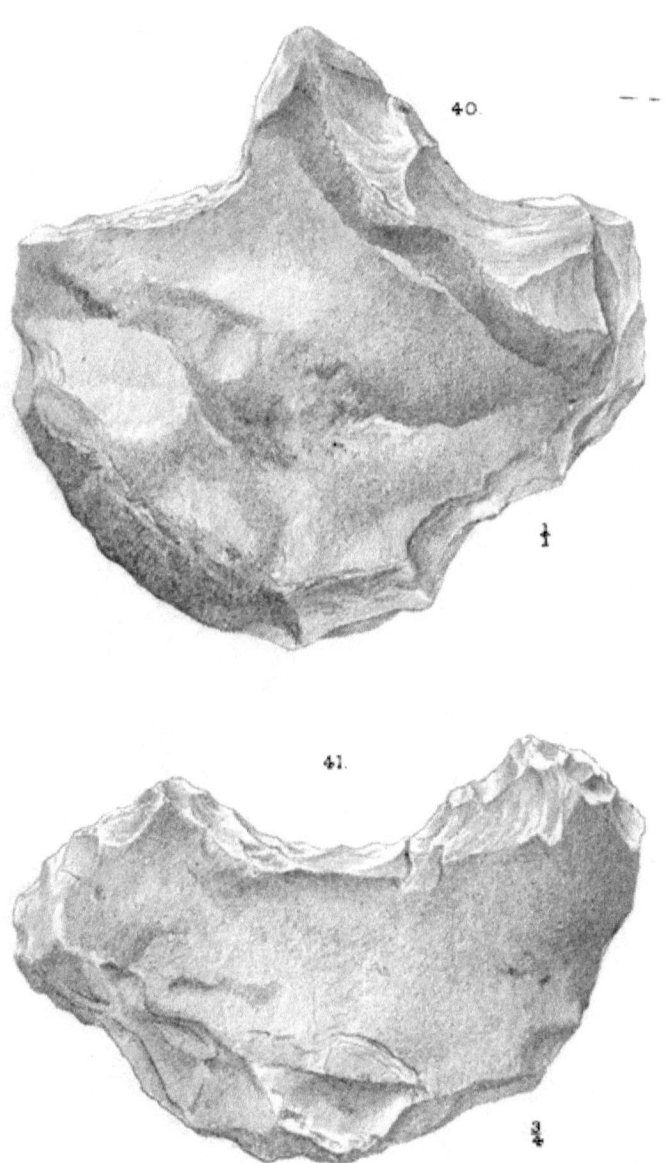

W S Tomkin, del & lith

IV

ON THE AGENCY OF WATER IN VOLCANIC ERUPTIONS, AND ON THE PRIMARY CAUSE OF VOLCANIC ACTION [1]

§ 1. Introductory Observations—The Vapour of Water considered as the Primary Cause of Volcanic Action

The important part played by water in Volcanic Eruptions is a well-recognised and established fact, but there is great difference of opinion among Geologists as to whether water should be considered the *primary or secondary agent*, and as to the mode, time, and place of its intervention. Whichever view may be adopted, it is evident that water plays an important part in the eruptions, and that the hydrogeological questions relating to the underground circulation and penetration of water have to be carefully considered. The objections to the chemical theory of Davy, according to which water finds its way to the interior of the earth, and there, meeting with the metals of the earths and alkalies, is decomposed with the evolution of intense heat, steam, and

[1] The general views expressed in this paper were laid before the Geological Section of the British Association at the York Meeting in 1881, and were published in a short abstract in the Reports of Section C, p. 610.

gases, have been so often stated, that it is not necessary here to refer to them further than to remark that the objections I shall have to urge generally against the percolation or passage of water to extreme depths will apply equally to this hypothesis also.

The theory of volcanic action which has of late years been most generally accepted is that of Mr. Poulett-Scrope. As formulated by him in the successive editions of his standard work on *Volcanos*,[1] "the main agent in all these stupendous phenomena —the power that breaks up the solid strata of the earth's surface, raises, through one of the fissures thus occasioned, a ponderous column of liquid mineral matter to the summit of a lofty mountain, and launches thence into the air, some thousand feet higher, with repeated explosions, jets of this matter and fragments of the rocks that obstruct its efforts— consists unquestionably in the expansive force of some elastic aëriform fluid struggling to escape from the interior of a subterranean body of lava, *i.e.* of mineral matter in a state either of fusion or at least of liquefaction at an intense temperature. This body of lava is evidently, at such times, in igneous *ebullition*."[2] He further explains that the rise of lava in a volcanic vent is occasioned "by the expansion of volumes of high pressure-steam generated in the interior of a mass of liquefied and heated mineral matter within or beneath the eruptive orifice," so that the vapour reaches the external

[1] *Volcanos*, by G. Poulett-Scrope, 2nd edit. (1862), p. 30.
[2] *Ibid.*, pp. 39-40.

"surface in a state of extreme condensation and entangled in the liquid lava which rises with and escapes outwardly, just as any other thick or viscid matter exposed to heat from beneath in a narrow-mouthed vessel *boils up* and *over* the lips of that aperture."

I might have felt some doubt as to the exact meaning of this passage, especially as the author proceeds to remark, that "at what depths those volumes of vapour are generated may be a question," but as he goes on to observe, " that the tendency to vaporisation must everywhere occasion an extreme tension" or expansive force throughout the mass, " only restrained by the enormous weight and cohesion of the superincumbent rocks," I infer from this and from the general tenor of his remarks that he considered steam to be the sole motive power throughout, and to exist in the molten magma itself. He says: " If any doubt should suggest itself whether this fluid is actually generated within the lava, or only rises through it, having its origin in some other substance, or in some other manner beneath, it must be dispelled by the evidence afforded in the extremely vesicular or cellular structure of very many erupted lavas, not merely near the surface, but throughout their mass, showing that the aëriform fluid in these cases certainly developed itself interstitially in every part. And although such vesicles or cells appear at first sight to be wanting in other lavas, at least in the lower portions of the lava-current after its consolidation, the microscope invariably, or almost invariably, discovers them. In those exceptional cases, where the rock is to appearance perfectly compact, it is allowable to suppose that the vapour it once contained escaped in ascending bubbles, or by exudation through extremely minute pores, or was condensed by pressure and refrigeration previously to the solidification of the matter."[1]

It must be borne in mind that for this hypothesis to have any value, the explosive material must extend throughout the mass of lava and act from its base upwards, just as much as it is necessary that the powder in the gun-barrel should be at the back of the shot. It should therefore extend to the volcanic foci, at whatever depth that might be, and be there occluded in the lava.

Professor Judd, in his summary of Mr. Scrope's views, remarks that on this hypothesis volcanic outbursts are considered to be "due to the accumulation of steam at volcanic centres, and that the tension of this imprisoned gas eventually overcomes the repressing forces which tend to its manifestation," and that "in the expansive force of great masses of imprisoned vapour, we have a competent cause for the production of fissures through which volcanic outbursts take place."[2]

Mr. Scrope does not enter upon the question of the mode in which the water has become occluded in the fluid magma. Sir Charles Lyell, however, in supporting the views of Mr. Scrope, says: "We may suppose that large subterranean cavities exist at the depth of some miles below the surface of the earth in which

---

[1] *Op. cit.*, pp. 36–7.
[2] Judd's *Volcanoes*, pp. 33 and 189.

melted lava accumulates, and when water containing the usual mixture of air penetrates into these, the steam thus generated may press upon the lava and force it up the duct of a volcano, in the same manner as a column of water is driven up the pipe of a geyser."[1]

Briefly the opinion of Mr. Scrope upon the cause of volcanic action is that it is to be attributed to the escape of *high-pressure steam* generated in the interior of the Earth. Before proceeding to discuss this hypothesis more fully, and to state my objections to the views of this distinguished volcanologist, I may refer to the paper of the late Mr. R. Mallet, in which the same subject is treated from an entirely different point of view.

According to Mr. Mallet, volcanic energy is *not the direct* product of primordial heat of fusion, although it is evidently due to the loss of that heat, and is the result of the cooling of our globe. He defines it thus: "The heat from which terrestrial volcanic energy is at present derived is produced locally within the solid shell of our globe by transformation of the mechanical work of compression or of crushing of portions of that shell, which compressions and crushings are themselves produced by the more rapid contraction, by cooling, of the hotter material of the nucleus beneath that shell, and the consequent more or less free descent of the shell by gravitation, the vertical work of which is resolved into tangential pressures and motion within the thickness of the shell.[2]

---

[1] *Principles of Geology*, 10th edit., vol. ii. p. 221.
[2] *Phil. Trans.*, vol. 163, p. 167.

The crushing of the earth's solid crust in this manner along lines of greatest weakness is considered by Mr. Mallet to evolve heat sufficient to fuse segments of the crust and to cause the extrusion of the fused masses. He says: "The result of the crushing is to produce irregular masses, on the whole tending to verticality, of pulverised rock, heated more or less highly, that may extend to any depth within the solid crust ; but it is only to such depths as water can percolate or infiltrate by capillarity that the deepest focus of volcanic activity can be found."

There are, however, many objections to Mr. Mallet's ingenious hypothesis. Amongst the most serious, and it seems to me fatal objections, are—(1) that it fails to explain the centralisation of the heat, for even though developed by sudden and paroxysmal movements, and confined to a narrow line, the heat would be dispersed and dissipated throughout the whole mass affected by the pressure ; (2) that there is an absence of volcanoes in the majority of mountain ranges where the pressure and crushing have been of the most powerful character ; (3) while, on the contrary, in many volcanic areas there is little or no evidence of great lateral pressure, or of much disturbance of the strata beyond fissuring.

The great mountain ranges of the Alps and Pyrenees, where the strata are tilted, contorted, and enormously crushed, do not contain a single volcano ; the strata are highly metamorphosed, yet show no traces of igneous fusion. In the Andes, the volcanoes are mostly situated on flanking ridges, or on the lower grounds at their base, and rarely on the high central ridges. We must, in fact, look for volcanoes on lines of fissure rather than on lines of fault. Lines of fault, even those of the greatest magnitude, show no fused walls, though the formation of slickenside surfaces must, seeing how powerful was the friction, have been attended with considerable heat.

I however entirely agree with Mr. Mallet in two of his propositions, namely, that volcanic action " is only one phase of a unique force which has always been in action .... since our planet was nebulous;" and in one sense "that without water we can have no volcano;"[1] that is to say, no explosive volcanoes.

§ 2. OBJECTIONS TO STEAM AS THE PRIMARY CAUSE OF VOLCANIC ERUPTIONS

The chief objections to the hypothesis that the vapour of water is the primary agent in volcanic action, are—

1. The absence of occluded water and steam in so many old and modern lava flows.
2. The insufficiency of the elevatory force.
3. The want of agreement in *time and proportion* between the discharge of *steam* and the discharge of *lava*.

On the hypothesis of Mr. Poulett-Scrope it has to be assumed either that the water formed part of the original molten magma, or that the surface-waters find a passage by percolation or by fissures through the crust of the earth to the molten volcanic foci beneath, and there become incorporated with the lava. That water does penetrate to considerable

[1] *Op. cit.*, p. 216.

depths is evident, and if nothing interfered to check its descent, the extent and range of the percolation would be comparatively unlimited. But various stratigraphical causes interfere with this transmission, such as impermeability of the strata, faults, and unconformity of superposition. These causes especially impede the flow of water in the more frequently disturbed Palæozoic strata, but the newer sedimentary strata are less affected by them. In Tertiary and Secondary strata water is always, or almost always, met with at most depths reached. It is otherwise with the Palæozoic strata generally, where, for example, as is so common in Coal Mines, the faulting of the strata divides them into separate sections or compartments, in each of which the supply of water from adjacent areas is cut off, and no fresh supplies being received from the surface, there is no further accession of water. Again, where Tertiary or Secondary strata overlie unconformably Palæozoic strata, the interstices on the surface of the older rocks are so effectually *plugged* by the basement bed of the superimposed strata, that they are often rendered perfectly watertight. There is a remarkable instance of this in the coal-field of Mons,[1] where the Coal-measures are in places worked under a depth of 600 to 900 feet of loose sandy Cretaceous strata full of water, and yet are found to be perfectly dry, owing to the circumstance that the basement bed of the overlying strata has formed a sort of puddle cover, sealing up, as it were, the edges of the underlying strata. The percolation of

[1] The author in *Proc. Inst. Civil Engineers*, vol. xxxvii. p. 129.

the surface-water to great depths is, in consequence of these interruptions, far from being so general as might be supposed.

Admitting, however, the possibility of water descending in certain cases—as through the fissures and crevices of crystalline rocks, or, in the absence of any special conditions to stop its descent, through permeable strata—it is a question whether its descent would not be stayed by the increase of heat at great depths.

It is known experimentally that the pressure of steam which at 212° F. equals one atmosphere, is, at a temperature of 432° F., equal to that of nearly 24 atmospheres; and also that the rate increases with the rise of temperature, and is faster at high than at low temperatures, as shown in the following table:[1]—

Temperature.	Pressure of mercury.	Pressure in atmospheres.	Temperature.	Pressure of mercury.	Pressure in atmospheres.
F.	Inches.		F.	Inches.	
212°	29·89	1·0	332°	216·21	7·21
222	36·35	1·21	342	247·38	8·25
232	43·91	1·47	352	281·99	9·40
242	52·72	1·75	362	320·28	10·68
252	62·92	2·10	372	362·50	12·08
262	74·69	2·50	382	408·92	13·63
272	88·18	2·94	392	459·80	15·33
282	103·58	3·45	402	515·41	17·18
292	121·08	4·04	412	576·02	19·20
302	140·88	4·70	422	641·90	21·40
312	163·18	5·44	432	713·32	23·88
322	188·22	6·27			

Thus, while an increase of 50° F. to the temperature of water vapour at 212° causes an increase of pressure of only 1½ atmospheres, the addition of another

[1] Balfour Stewart's *Elementary Treatise on Heat*, 2nd edit. Table I., p. 392.

50° gives an additional pressure of nearly 3 atmospheres; another 50° gives 5 atmospheres; while 50° more gives a further increase of 8½ atmospheres.

This rate of increase is nearly proportionate to the 5th power of the excess of temperature above −40° F. Pouillet employed empirical formulæ to ascertain the probable pressure at higher temperatures up to 516° C. These, although not mathematically exact, are sufficiently so for our purpose—at all events up to the temperature of 773° F., the critical point of water, at which new conditions set in.

Temperature.	Pressure in atmospheres.
510° Fahr.	50 atmospheres
592	100 ,,
686	200 ,,
747	300 ,,
794	400 ,,
832	500 ,,
864	600 ,,
893	700 ,,
918	800 ,,
942	900 ,,
962	1,000 ,,

At the critical point of water the steam pressure would amount to nearly 350 atmospheres, or, if the rule holds, to the temperature at which water may undergo dissociation, we should then have a pressure exceeding 1,000 atmospheres. In any case these conditions point to a possible term at which the expansive force of vapour will exceed that of the hydrostatic and rock pressure and the descent of the surface-waters is in all probability stayed.[1]

[1] M. Delesse, in his paper on the water in the interior of the Earth, considered that notwithstanding that water tends to pass

Adopting the mean thermometric gradient of 48 feet of depth per 1° F., assumed in the paper (No. VI.) on Underground Temperature, the following will be the relation between depth and temperature down to a depth of 150,000 feet or 28½ miles, taking a mean annual surface temperature of 50° F. :—[1]

Table of Temperatures at Depths with a Thermometric Gradient of 48 feet per 1° F.

Depth.		Temperature.	
Surface		= 50° F.	
500	feet	= 60½	
1,000	,,	= 71	
1,500	,,	= 81	
2,000	,,	= 92	
3,000	,,	= 112	
4,000	,,	= 133	
5,000	,,	= 154	
7,776	,,	= 212	Boiling point.
10,000	,,	= 258	
15,000	,,	= 362	
20,000	,,	= 467	
30,000	,,	= 625	
34,704	,,	= 773	Critical point.
40,000	,,	= 883	
50,000	,,	= 1,092	
100,000	,,	= 2,133	
150,000	,,	= 3,175	

into vapour at the high temperature of great depths, the pressure of the overlying strata, and the resistance they there offer to its return being greater than its tension, would cause it to retain its liquid state. But at a depth, which he estimates at about 60,000 feet, and at a temperature of about 1,100° F., the overlying pressure (taken at the rock weight), and the elastic force of the vapour of water would be in equilibrium.—*Bull. Soc. Géol. de France*, 2nd ser., vol. xix. (1861), p. 64.

[1] Or the rate may be calculated separately for the three geological areas with their different thermic gradients (see Paper No. VI.).

Or if in districts of crystalline rocks and slates, and with a gradient of 45 feet, a temperature of 3,000° might be reached at a depth of about 26 miles.

Besides these, there are other considerations which should not be overlooked, although it is impossible at present to assign a value to them. One is whether there may not be areas of certain rocks in which the gradient is more rapid than in other areas; and another whether in tropical regions there is not generally a more rapid thermometric gradient. In the paper on Underground Temperatures, a few of the observations raise these points as questions for further inquiry.

Another point, which I have before mooted,[1] is more purely hypothetical. It is whether the effect of the excessive cold of the Glacial period—a cold prolonged during so many thousands of years—may not possibly have left its mark on that portion of the earth covered for so long a period by perpetual snow and ice—whether the loss of heat in the upper layers of the crust may not only have altered the thermometric gradient, but also induced, as it were, premature contraction by an abnormal abstraction of heat during that period. Whether also the outer portion of the crust so affected might not now present a lesser gradient than the present mean surface temperature would warrant, and whether or not this might possibly be an element in the present effective rigidity of the crust.

Taking into consideration all the conditions to

[1] *Phil. Trans.*, vol. 164, p. 305.

which water becomes subject with increasing depth and temperature, together with the circumstance that while the pressure of Water increases with depth in *simple arithmetical progression,* that of the elastic vapour of water is one of a very *rapid geometrical progression,* it becomes extremely improbable that water can penetrate beyond a certain depth beneath the surface. Roughly, it is a question whether 6 to 7 miles would not be a limit. At all events I feel it impossible to accept any hypothesis based upon an assumed percolation to unlimited depths.

It is true that the experiments of Daubrée, which will be further alluded to, show that owing to the force of capillarity, water can pass through porous strata against a considerable resisting pressure, but on the other hand Wolff's experiments show that the effects of capillarity decrease with the increase of temperature, and tend to prove that there is a point at which they would altogether cease.

It may also be a question whether at the high temperature at great depths the vapour of water would not undergo decomposition, for M. H. St. Claire Deville[1] has shown that under certain conditions, at a temperature of from 1,103° to 1,300° C., it is dissociated into its elements, and in so dissociating it augments its volume by one half, and increases its pressure in proportion.[2] That water is

[1] "Sur le phénomène de la dissociation de l'Eau," *Comptes Rendus,* vol. lvi. p. 195.

[2] At the same time it is to be observed, that enclosed in a platinum tube water did not decompose at a temperature near the fusing point of the platinum.

decomposed in contact with lava during eruptions is rendered probable by the investigations of M. Fouqué during the last great eruption of Santorin, for he found that the gases given off under water during the eruption, and collected as they ascended through the sea, often contained as much as 30 per cent. of free hydrogen. From this circumstance, and the fact that he also found free oxygen occasionally present, he considered it not unlikely that the vapour of water exists in a state of dissociation in the lava during eruptions.[1]

Other geologists have contended for the possibility of water gaining access to the volcanic foci by fissures opening into the sea-bed,[2] whilst the fissures are supposed to be formed by the molten matter struggling to escape. To this it has been rightly objected, that in such a case the lava would at once fill the fissure to the exclusion of the water. By others it has been suggested that the fissures are caused by the escape of imprisoned elastic vapours; but, as Mr. Scrope remarks, this is reasoning in a circle, for while it supposes the aqueous vapour to be the cause of the disturbance, it introduces the water *after* the effects attributed to it had been produced.

The further objection is, that it would not be possible for steam alone, even at its highest pressure, to force forward and *gradually* erupt the column of lava in the duct of the volcano. Bischof's[3]

---

[1] *Santorin et ses Éruptions*, p. 232.

[2] M. Angelot, *Bull. Géol. Soc. de France*, vol. xiii. p. 178; vol. xvi. p. 43; and 2nd ser., vol. i. p. 23.

[3] *Edin. New Phil. Journ.*, vol. xxvi. (1839), p. 132.

hypothesis was founded on an erroneous estimate of the elastic force of steam.

But the objection to which I attach most weight and importance is one which deals with facts which are within the field of actual observation. On the hypothesis that attributes the extrusion of lava to "the expansive force of some elastic aeriform fluid, struggling to escape from the interior of a subterranean body of lava," it would follow that no lava could escape without the accompaniment of that elastic aeriform fluid, nor could any large evolution of vapour or gases take place without a corresponding escape of lava, for the relative discharge of steam and lava would necessarily bear some relative proportion one to the other.

§ 3. Facts not in Accordance with the Hypothesis

Although the phenomena accompanying volcanic eruptions are so constantly recorded, those which bear in particular on this question are generally so mixed up with the other details, that it is not always possible to determine their relative bearing and sequence. Still there are an ample number of carefully recorded cases to show that the discharge of lava is not in proportion to the discharge of steam, nor is the discharge of steam always in accordance with the escape of lava, which they should be if the hypothesis were correct. These conditions would on the contrary seem to be perfectly independent one of the other. It is of course conceivable that paroxysmal explosions may disperse the lava in aerial discharges, and reduce the volume of the outflow; but these would not affect the more general results. There are too many great eruptions that have been attended with a small discharge of lava, and too many of the largest lava streams have been erupted quietly and with a very small exhibition of explosive violence, to allow of much doubt on the subject. Sometimes, when the discharge of lava has been at its maximum, the explosive violence has been at its minimum, and, on the other hand, violent detonations have been attended with small overflows of lava.

According to Daubeny,[1] there is no recorded lava-flow accompanying the eruptions of Vesuvius prior to the eruption of A.D. 1036. This, however, may be the mere absence of record. Still, it would seem to point to the prevalence of paroxysmal eruptions like that of the great eruption of 79 B.C., rather than of any important discharge of lava-flows.

Mr. Scrope divides volcanic outbreaks into periods (1) of moderate activity, and (2) of paroxysmal violence,[2] and he himself remarks that "the volume of lava poured out by an eruption does not preserve any constant proportion to the force or continuance of its explosions," but he offers no explanation of the fact. He instances Etna[3] as an example of almost continual moderate activity with occasionally more or less paroxysmal outbursts. The volcano of the Island of Bourbon offers another example of the same kind. He also points out[4] that "in all cases

---

[1] *Description of Volcanos*, 2nd edit., 1848, p. 225.
[2] *Volcanos*, 2nd edit., pp. 16–19.
[3] *Ibid.*, p. 24.   [4] *Op. cit.*, p. 23.

where lava is emitted its protrusion marks the crisis of the eruption, which usually attains the maximum of its violence a day or two after its commencement. The stoppage of the lava in the same manner indicates the termination of the crisis, but not of the eruption, for the gaseous explosions continue often for some time with immense and scarcely diminished energy." Vesuvius "has often continued in eruption for periods of several months, discharging moderate jets of scoriæ, lapilli, and sand from temporary orifices at the summit or flank of the cone, or at the bottom of its crater, when there was a crater; while streams of lava welled out, *sometimes almost with the tranquillity of a water-spring* from the same or from contiguous openings." [1]

Professor Palmieri [2] says of Vesuvius, that on some occasions the eruptions commence with explosions and detonations of greater or lesser violence, ending with a great eruption and a copious flow of lava; and that at other times great eruptions have taken place without any precursory signs.

Professor Phillips observes of the great eruption of Vesuvius of 1794, which was characterised by the flow of some of the largest lava currents ever erupted from this mountain, that "for nearly a month after the eruption (of lava), vast quantities of fine white ashes mixed with volumes of steam were thrown out from the crater." [3]

---

[1] *Op. cit.*, p. 17. The italics here and in the following pages of this chapter are mine.—J. P.

[2] *Eruption of Vesuvius of* 1871–2. Mallet's Translation, pp. 94, 99–100.   [3] *Vesuvius*, pp. 92–4.

M. Ch. St. Claire Deville[1] states that the great eruption of Vesuvius in 1855 was *one of the most tranquil*. The projections only lasted a few days, and the detonations soon ceased. The lava continued to flow for twenty-eight days, and formed the largest current which has passed out in the north-west direction.[2] This eruption was in great contrast with that of 1850, which was one of the most violent and paroxysmal, when the mountain was changed in form, the central cone reduced, and the crater enlarged to 2 miles in circumference, *yet the flow of lava was comparatively small*.

The eruption of Etna of 1852 was one of unusual magnitude, and the flow of lava greater than ever witnessed, except probably in 1669. It commenced in August with violent explosions and ejection of scoriæ. The lava then began to flow from several openings, and flooded the country for a length of 6 miles and a breadth, in places, of 2 miles. The ejections of scoriæ continued during sixteen days, but *after that time they almost ceased*, except in a few smaller craters, though dense volumes of steam were occasionally discharged from the central crater, *but the flow of lava continued with little interruption through September, October, November, and December, and did not entirely cease until May,* 1853.[3]

An eruption, which seems of itself almost sufficient to prove the independence of the forces causing the ex-

---

[1] *Bull. Soc. Géol. de France*, 2nd ser., vol. xii. p. 1065.
[2] *Vesuvius*, p. 107.
[3] Lyell, *Phil. Trans.*, vol. 148, p. 18.

trusion of lava, and of those connected with the elastic vapours, is that of Santorin in 1866, recorded by M. Fouqué.[1] In the centre of the bay, enclosed by the great encircling old crater-walls now forming the islands of Thera and Asprosini, stands the small island of Kaimeni, the product of later eruptions. On the 26th January, 1866, the loose blocks on the southern slope of this island began to move, and on the 27th slight shocks were felt, gases evolved, and fissures rent in buildings. The ground in a small sandy bay was observed to rise, and by the 4th February the erupted mass, which consisted of blocks of lava, had attained a height of 32 feet. By the 5th, this protuberant mass of lava had increased to 230 feet in length by 98 feet in width, and 65 feet in height, and on the 7th to $492' \times 197' \times 98'$. The adjacent water was hot and the surface of the lava was consolidated, though it was incandescent at night. Until the 12th February there were no detonations and no explosion, notwithstanding the large quantity of lava emitted, for it was not confined to the matter above water, but it was, in places, gradually filling up the bay itself; and where there previously had been soundings of 103 fathoms, the depth was now reduced to from 40 to 70 fathoms.[2]

M. Fouqué remarks,[3] that "at the beginning of the eruption the discharge of lava was the most salient phenomenon; *the rock-emission proceeded in silence;* it was only at the end of several days that the explosions and ejections commenced and a crater

---

[1] *Santorin et ses Éruptions*, Paris, 1879.
[2] *Ibid.*, pp. 36 *et seq.*   [3] *Ibid.*, p. xv.

formed (the volcano of Giorgios)." The explosions attained great violence on the 20th and 22nd, and on the latter day the column of vapour and ashes rose to a height of about 7,000 feet. In April and May lava flowed more freely. The eruption was prolonged to 1869, when *the explosions were still frequent but the discharge of lava very small.*

Another eruption commenced in February, 1867, in the sea-bed west of Kaimeni, and by the 17th an island (Aphroessa) was formed 328 feet long by 196 feet wide and 32 feet high; while the adjacent sea-bed was in places reduced from a depth of 296 fathoms to 108 fathoms. This also was effected *quietly and without noise,* and it was not until later that the explosions began.

So noiseless and so steadily continuous was the protrusion of these masses of lava at first, that Dr. Cigalli, who watched them from day to day, compared their growth to the steady and uninterrupted growth of a soap-bubble.

Much of the lava of this great eruption was very compact, and not at all scoriaceous.[1]

Equally remarkable for its magnitude, and at the same time for its quiet, was the eruption of Mauna Loa in 1855. In speaking of this eruption Dana says that there *was no earthquake, no internal thunderings, no premonitions at the base of the mountain.* A small glowing point was seen at a height of 12,000 feet which gradually expanded, throwing off coruscations of light. A vent or fissure then formed, from which a vast body of

[1] *Op. cit.,* p. 72.

liquid lava *rapidly but quietly flowed during several weeks* (a later account says 10 months), forming a stream of lava which extended a distance of 65 miles, with a breadth of from 3 to 10 miles. He adds that those eruptions of fiery cinders which mark so strikingly Vesuvius, were almost wanting about the craters and eruptions of Mauna Loa, and the few that there were, were mainly in connection with the lateral cones.

On the other hand, Mr. Scrope remarks that the great paroxysmal eruptions of volcanoes " begin generally with one tremendous burst, which appears to shake the mountain from its foundations. Explosions of aeriform fluids, each producing a low detonation and gradually increasing in violence, succeed one another with great rapidity from the orifice of eruption, which is in most instances the central vent or crater of the mountain."[1]

One of the most violent of the explosive eruptions was that of Cosequina in 1835.[2] This volcano is situated on a promontory south of the Bay of Formosa in Central America. The detonations were so violent that they were heard at a distance of 280 miles. So enormous was the quantity of ashes and scoriæ shot out of the crater, that for a distance of 25 miles they covered the ground to a depth of about 15 feet, and the finer dust was carried by the wind as far as Jamaica, a distance of 800 miles. It is not recorded that this great outbreak was accompanied by any

[1] *Volcanos*, 2nd edit., pp. 20–21.

[2] The great eruption of Krakatoa has taken place since this was written. It was one of the same character; and has been investigated by a committee of the Royal Society. (*Trans.*, 1888.) The cause and origin of the eruption did not come under discussion.

lava-flow.[1] The mountain itself rises only 480 feet above the sea-level.

From time to time the violence of the paroxysmal eruptions has blown off and truncated the cone of various volcanoes, and enlarged their craters, from the small dimensions they had when the eruption issues at the built up apex, to gulfs sometimes several miles in circumference and of great depth, eviscerating, as it were, the very centre of the mountain. Scrope[2] mentions as examples of such paroxysmal eruptions —13 eruptions of Vesuvius, 8 of Etna, 2 of Teneriffe, 1 of San Georgis in the Azores, 3 of Palma, and 1 of Lancerote (Canary Islands), and all the recorded eruptions of Iceland.

"Sometimes in these eruptions no absolute escape of lava takes place, scoriæ alone being projected. In all cases when lava is emitted its protrusion marks the crisis of the eruption, which usually attains a maximum of its violence a day or two after its commencement. *The stopping of the lava* in the same manner indicates the termination of the crisis, but not of the eruption, *for the gaseous explosions continue often for some time with immense and scarcely diminished energy.*"[3]

§ 4. THE INSUFFICIENCY OF OCCLUDED VAPOUR

It seems therefore evident that there is *no definite relation between the quantity of explosive gases and vapours and the quantity of lava* discharged from the volcanic foci. It is conceivable that the enormous

[1] Reclus, *La Terre*, p. 668. [2] *Volcanos*, p. 25.
[3] *Ibid.*, p. 23.

force of some of the explosions may, in the paroxysmal outbursts, shatter and blow to fragments the lava as it rises in the crater, but this appears hardly sufficient to account for the want of a quantity in some proportion to the vast volumes of vapours, were those vapours the cause of the extrusion of the lava. It is still more difficult to conceive on this hypothesis the frequent vast discharge of lava in tranquil eruptions without a greater escape of vapour.

If, as the advocates of that theory maintain, the escape of lava depended on the escape of the imprisoned vapours, it is not easy to see how the constant supply, whether of the lava or of the steam, is maintained. The rise and escape outwardly of the lava in a volcanic vent has been likened to the boiling up and over of any other thick and viscid matter exposed to heat from beneath in a narrow-mouthed vessel,[1] and Constant Prevost compared it to the overflow of a liquid caused during fermentation by the evolution of carbonic acid gas. But the cases are not analogous. In the one instance the boiling up is caused by the presence of a substance in which water is a constituent part. In the other, the evolution of gas arises from the decomposition of the substance affected, which is not the case with lava.

We have already pointed out the difficulty of accounting for the introduction of water into the deep-seated volcanic foci. Even supposing it to have been introduced so as to cause a boiling over, that ebullition would go on so long as any of the imprisoned vapour remained in the lava; but when its

[1] Scrope's *Volcanos*, p. 40, and Lyell's *Principles*, vol. ii. p. 221.

expulsion was effected, the introduction of fresh supplies from outside would, as in the case of the water in the Geyser pipes, become necessary, or the boiling over of the lava would cease. If, on the contrary, the water were present in combination with the lava in the volcanic foci, there is no reason why the passage to the exterior once opened the eruption should cease until all the mass susceptible of boiling over should be expelled, in which case the volcanic activity would be indefinitely prolonged and then become extinct after one eruption.

The only logical hypothesis on which I can conceive the vapour of water to be present in the fluid magma of the volcano is the one suggested by Dr. Sterry Hunt, who considered that the magma is not part of the original molten anhydrous nucleus of the earth, but an intermediate layer derived from the first outer crust of old surface rocks which had been exposed to meteorological agencies, and retained, when fused under pressure, the water with which they had become permeated when on the surface. He supposes the original central nucleus to have gradually become solid by pressure and loss of heat, while, at the same time, an outer crust formed. As that crust became thicker and covered by sedimentary strata accumulated upon it, its under surface, owing to the rise of the isothermal bars, was gradually remelted, thus forming an intermediate fluid layer between the solid nucleus and the solid outer crust.

Mr. O. Fisher, on the other hand, considered that the interior of the earth has shrunk more than mere cooling alone would account for, and suggested that this was due to the presence of superheated water in large quantities in the original nucleus, and that the blowing off of this water during volcanic eruptions might have contributed materially to the diminution of the volume of the magma.[1] In a subsequent work [2] Mr. Fisher applies this hypothesis more particularly to the explanation of volcanic action. He supposes a solid crust of about 25 miles thick resting on a fluid substratum of highly heated rocky matter in a state of igneo-aqueous fusion, and states that if a crack were produced by any cause in the under surface of the crust it would become filled with the water-substance or vapour given off from the fluid magma at a high tension. Whenever the rent, commencing below, opens upwards, vapour at a high tension will escape, and after a certain time will be followed by the magma itself, which will overflow at the surface because the water-substance expanding, owing to the diminished pressure, would render the whole column of less weight than an equal column of the crust. On this view he considers that any disruption in the crust which is sufficient to permit the passage of steam at an enormous pressure would originate a volcano; and "much of the lava poured out might consist of the materials of the crust itself, fused by the passage of the gases through it, and so vary in its composition at different vents, and even at the same vent at different times."

I need not dwell on several objections I feel to both these hypotheses, because the one just mentioned

---

[1] *Trans. Cambridge Phil. Soc.*, vol. xii. p. 414.
[2] *Physics of the Earth's Crust*, chap. xv. pp. 185 *et seq.*

applies equally to them—namely, that, if they were true, all rocks formed under such conditions should exhibit evidence of the presence or of the escape of vapour, while also in the latter case the presence of water in the magma remains unaccounted for. According to these hypotheses, all volcanic matter should be more or less scoriaceous, whereas there are many lavas which are little, and others not at all, scoriaceous. The great sheets of basaltic rocks which have welled out from fissures at former geological periods are, as a rule, neither scoriaceous, unless very superficially, nor are they accompanied by *débris* indicating explosions and projections due to the escape of vapour and gases. Why also should not all rocks of igneous origin, as well as volcanic rocks proper, be scoriaceous, if such were the conditions of the molten magma beneath the solid crust? The general want of hydration in volcanic rocks and their associated minerals is likewise incompatible with the assumption.

It has been suggested by some writers that large subterranean cavities may exist at depths in the earth's crust, and that the vapour of water under high pressure is stored up in such underground cavities. But the pressure of the strata is so great at depths, that,—as in deep coalpits where no permanent cavities can be formed owing to the "creeping" of the strata,—it would be impossible for such cavities to exist in Sedimentary strata, while in Igneous rocks the initial plasticity of the rock and pressure would be obstacles to the formation of any cavities. Even if such cavities did exist, they could only be maintained by the action of an elastic fluid, whose pressure would exceed that of the superincumbent strata. Geology affords no evidence of such underground reservoirs, or of any having existed in former times. No explosions of steam, except from surface waters, show themselves during the disturbances, shocks, and rents which accompany earthquake movements; nor does any persistent issue of pent-up steam give countenance to the supposition that the water permeates the rocks to great depths or exists there in natural cavities.

Natural cavities at depths in the earth's crust I hold to be impossible. There may be cavities in the Igneous rocks near the surface, due either to contraction, to rapid cooling, or to the shell left by the escaping lava streams. But these cannot take place at great depths. They are all connected with subaerial conditions.

With regard to such cavities as those so common and of such extent in limestone rocks, it must be remembered that these cavities are entirely due to the descent of the surface waters to a definite underground level, and to their escape along lines of fissure, either into the adjacent valleys, or at the tide-line on the adjacent coast. Below those levels there can be no active circulation of water, and no possibility, therefore, of great cavities, due to the passage of water through underground channels, being formed. Changes of level may have carried some of these cavities to certain depths beneath the surface, but that they should have been carried to the great depths we are referring to, or be able to withstand the pressure, is more than problematical. In limestone strata they occur near the surface, or at a short distance beneath the surface. Wherever these rocks have been worked at a depth beneath the line of water saturation, such cavities are of very rare occurrence. Deep mines reveal occasionally a few fissures, and some comparatively small cavities, but these are mostly in mineral veins, which show no relation with volcanic phenomena.

§ 5. Influence of Volcanic Eruptions on Spring and Well Waters

It is a singular circumstance that although the presence of water in volcanic eruptions has been so long recognised, and the disturbances suffered by wells and springs have been so often noticed, no systematic series of observations has been made, either on the surface or on the underground waters, in connection therewith. There are many allusions and incidental notices, but nothing in the form of special and exact details. Most writers on the subject speak of the disturbances to wells and springs as a common or obvious fact; but a series of extended and accurate observations is much needed.[1] In the absence of more exact data, we have to avail ourselves of the few local observations made by witnesses who happened to be on the spot.

The eruption of Vesuvius of 1813–14, which commenced with a few trifling explosions and shocks in September, and by some small eruptions of lava in October and November, followed by the

[1] The observations should not be limited to the volcanic area, but should extend to the Sedimentary strata at a distance.

great eruption of December, was witnessed by M. Menard de la Groye,[1] who remarks that "towards the end of May the well-waters of Torre del Greco and Torre dell'Annunziata failed, and that this was an ordinary precursory symptom of the eruptions." In June the waters continued further to lower, and "in the first fortnight in July they fell so low as to alarm the population," while "in October the wells of Resina, Torre del Greco, and other places failed in a surprising manner."

Professor Phillips briefly records[2] the following instances :—" July, 1804. Severe earthquake— diminution of springs." In May, 1812, the wells failed or were much lowered at Torre del Greco and Resina, as well as a thermal spring. In June, July, and August heavy rains occurred; yet this did not restore the water in the wells, which still remained low, and even lower than before in September, and this scarcity was felt along the whole Vesuvian coast, and in the valley of the Sarno. Early in 1822 the wells lost their water. August, 1833, water failed in the wells.[3] The loss of water has sometimes been attributed to other causes, such as the state of the rainfall, &c., but Professor Phillips specially observes that this sinking of the wells cannot be explained by reference to the previous state of the weather;[4] and, after a careful examination into all the phenomena connected with the eruptions of Vesuvius, he alludes again to "the sinking of water in the wells

[1] *Journ. de Phys. et de Chim.*, vol. lxxx. p. 390.
[2] *Vesuvius*, pp. 96 et seq.   [3] *Ibid.*, p. 140.
[4] *Ibid.*, p. 141.

around Vesuvius—the total drying up of some and the increased descent of the bucket in all," during times of Volcanic disturbances, as an important fact.

M. Ch. St. Claire Deville[1] remarks: "It is well known that there is only one tolerably certain indication of an approaching eruption of Vesuvius, and that is the disappearance of the water in the wells of Resina and Torre del Greco."

According to Poulett-Scrope,[2] the threatening indications of an approaching crisis "are accompanied by the disturbance or total disappearance of springs, and such accidents as the cracking, splitting, and heaving of the substructure of the mountain must naturally occasion."

Professor Guiscardi, of Naples, in answer to my inquiry, wrote:[3] "As a rule the water of wells in the neighbourhood of Vesuvius undergoes changes in quantity, and even quite disappear before the commencement of eruptions. Only as well as I know in the eruption of 1861, the phenomena followed the eruption. I add a list of such diminishing and drying wells."

"1843. Decrease of water in the wells of Resina; it was preceded by emission of lava.
"1846. Some wells of Resina dried, and emission of lava followed.
"1846. Six adventive cones in the crater; water decreases at Resina in wells.
"1847. Decrease of water in the wells of Resina; great lava flowings.

---

[1] *Bull. Soc. Géol. de France*, 2nd ser., vol. xiv. p. 254 (1856).
[2] *Volcanos*, 2nd edit., 1862, p. 21.
[3] Letter to the author dated 1st September, 1881.

"1848. Water decreases in the wells of Resina and Torre del Greco. Earthquakes in the neighbourhood of Vesuvius. Lava flowing.

"1849. The same decrease of water—strong explosions, bellowing, and lavas.

"1850. January 23rd, at Resina and Torre del Greco decrease of water in wells. Strong explosions. February 5th, lava poured out with bellowing.

"Before the eruption of 1794 there was at Torre del Greco a small torrent, capable, it is said, of moving four mills. After the eruption the torrent got very poor, so that the water scarcely supplied a fountain. After the eruption of 1861, there was an increase of water in this fountain and in some small springs near the shore, and one was noticed in the sea itself, which lasted nearly a month."

"There is a well on a farm of some relatives of mine (Guiscardi's), at St. Georgio di Cremano, in right line nearly 4 miles from Vesuvius, 150 feet deep, and plentifully fed by a spring. After the eruption of 1861 the water began to decrease, and a year after it was quite dry. This was followed by so abundant an emission of carbonic acid, that the well had to be stopped up."

I must observe, however, that the high authority of Professor Palmieri does not favour this view.[1] He states that previous to the eruption of Vesuvius in 1871–72, the water in the wells was neither deficient nor scarce, but was very acid afterwards. He elsewhere mentions that he considers these supposed premonitory signs either only to happen occasion-

[1] *The Eruption of Vesuvius of* 1872, translated by R. Mallet, F.R.S., p. 135.

ally, or to be mere coincidences, such as the coincidence of a dry or rainy season. But the weight of evidence is certainly against this opinion, and, as I shall presently explain, there may be tracts that have an independent water-level which escape the surrounding disturbance, and it is not impossible that this very circumstance has led to the selection of such areas for the sites of towns and villages on the slopes of the mountain.

The line of underground water-level is subject to great fluctuations. It varies with the rainfall and with the eruptions of the Volcano. This variation of level is shown in the diagram of Etna, p. 124. There were eruptions of Vesuvius in 1865 and 1867, but none in 1866, and though there were a number of minor shocks, no important earthquake is recorded in that year.

The more local springs which supply the shallow surface wells may remain undisturbed, while at other points the deeper-seated springs having a wider range may be tapped and drained. Again, the water in the superficial volcanic strata may flow outwards towards the circumference of the mountain, in consequence of the beds by which they are held up dipping from the central crater; while the springs in the underlying Sedimentary strata may dip towards and discharge into the volcanic duct. Some irregularity in the phenomena is therefore to be expected, and though in most places the wells suffer, it is quite intelligible that in others they may be but little affected.

The evidence of M. Mauget, a well-engineer of

great experience, respecting the sudden changes of water-level in the Neapolitan area, during his residence there, is important. He says that in May, 1866, the wells and springs around Naples began to be affected, and continued to diminish until June, but he considered that this might be due to ordinary causes, such as lesser rainfall. On the 29th June a sudden change took place. The waters of the aqueduct, which brought in the water from a distance of 12 miles, and of the canal of Lagno di Mofita, as well as of various rivers, became troubled and reduced in a surprising manner. The next day the waters became bright, but were found to be reduced to the extent of one fifth of their volume. The great springs of the Sannio district were reduced by one third; and the town of Sorrento was deprived of all potable water. This water is brought in by an aqueduct from the neighbouring hills, which consist of Eocene or of Cretaceous strata. The whole district, from the foot of the Apennines to the Neapolitan coast, was affected over an area of above 60 miles square.

At the same time various artesian wells in the valley of Sebito became sanded up and greatly reduced in their flow, and the two deep artesian wells of Naples threw up above 200 cubic mètres of trachytic and pumiceous sands and lapilli.[1]

It is evident that this diminution in the surface waters could only have been caused by their absorp-

---

[1] "Sur les variations subites dans le régime de divers cours d'eau dans l'Italie Méridionale," *Comptes Rendus*, vol. lxiv. p. 189 (1867).

tion under ground, either to restore some water-level reduced by a former eruption, or to fill fissures in course of formation preceding the eruptions of 1867 and 1868.

It seems to me, therefore, to use the words of Professor Phillips, that the observations respecting the effects produced on wells and springs by volcanic eruptions and earthquakes "have been too often and too carefully made to allow of a serious doubt on the subject." He asks, "What is the cause of it? and why is it an indication of coming disaster?"

§6. The Hydro-geological and Statical Condition of the Underground Waters in and under a Volcanic Mountain in a State of Rest

The cause is, I believe, not far to seek, when the hydro-geological conditions of the strata composed above of volcanic materials, and below, of Sedimentary strata, are considered.

So well known is the absorbent power of a volcanic surface, that the mention of the fact hardly seems necessary. On ordinary strata it is roughly estimated that about one third of the rainfall passes under ground, but on volcanic surfaces the whole rainfall soon disappears, a small proportion only being lost by evaporation. Amongst innumerable notices of this fact, it will suffice to mention those of two experienced authorities. Lyell remarks on the dry and arid surfaces of Etna, and on the rapid absorption of the rainfall, and observes that "the volume of rain-water and melted snow commonly absorbed by a lofty mountain like Etna is enormous;[1] again, Piazzi Smyth, in describing his ascent of Teneriffe, says, "that though so much rain had fallen lately, not a trickling stream, not even a drop of standing water, was anywhere to be seen; the pumicestone ashes had swallowed all up."[2]

Volcanic mountains being composed of streams of lava of very variable width and length, irregularly alternating with more widely spread layers of scoriæ and ashes, the whole mass would be permeable were it not that the decomposition of some and the consolidation of other beds, by atmospheric and aqueous agencies, have formed here and there impermeable beds, which intercept the rain-waters, and furnish local supplies to wells and springs (see Fig. 2). But where such impermeable beds do not intervene, the rain-water penetrates to depths, and is there stored until the line of water-level rises to such a height that the hydrostatic pressure forces it outwards, and causes it to escape as springs either temporary or perennial (*a*, *b*, Fig. 1).

Solid lava is impermeable, but water penetrates through, and is held in, the numerous fissures and cavities by which it is traversed. These fissures are due to contraction resulting from the cooling of the lava, and to fractures produced by subsequent disturbances; whilst larger cavities are produced by other causes. Of these, the two most important are—

First, the escape of vapours while the lava is con-

[1] *Phil. Trans.*, vol. 148 (1858), p. 763.   [2] *Teneriffe*, p. 349.

solidating. Sometimes the hardened outer crust of the lava is raised in great blisters, which, on the escape of the vapour, are sufficiently solid to retain their position, and remain like so many empty beehives on the surface of the lava streams. The Grotta delle Palombe, on Etna, which, according to Waltenhausen, has a length of about 500 feet, and a height in places of from 70 to 80 feet, and the great ice cave near the top of the Peak of Teneriffe, described by Piazzi Smyth,[1] and so large as to contain a lake of water of considerable size, are attributed by them to the escape of elastic vapours. They seem to me, however, to belong more probably to the following category.

Second, to the escape of lava from a lava stream after the exterior of it has become solid, when an empty shell in the form of a cave or tunnel is left. These tunnels or caverns are of not unfrequent occurrence, and are sometimes of large size. Scrope observes[2] that "among the lavas of Etna, Bourbon, Iceland, St. Michael, Teneriffe, and many others, caverns of very large dimensions are thus formed beneath the surface of a lava stream, and often imitate in their extent and windings the well-known caves worn by water in limestone rocks." Phillips and others notice the occurrence of similar tunnels in the lavas of Vesuvius, but they are all small size.

In the great Volcanic mountains of South and Central America, Humboldt long ago inferred that large cavities filled with water must exist in consequence of the ejection of water, with small fishes

[1] *Teneriffe*, p. 352. [2] *Volcanos*, 2nd edit. p. 79.

and tufaceous mud, from fissures caused by the earthquake shocks which precede the eruptions of the Volcanoes in the Andes.[1]

A French geologist, M. Virlet d'Aoust, has, moreover, given particulars of two great tunnel-caverns of Central America,[2] which will serve to indicate the magnitude of some of these subterranean reservoirs. The first is that known as the *Cueva de Chiuacamoté*, near Pérota, which he was assured extended several leagues in length (!) He found it to be a cavern of great size, and divided into compartments by falls of the roof. The floor is covered with a sandy gravel, and the side walls, here as in the cave of Custodio, exhibit grooved lines covered with slight calcareous incrustations indicative of old water-levels. The other is the *Breña de Custodio*, in the State of San-Luis Potosi, of which he says that it forms a perfect semispherical tunnel of the size of our largest railway tunnels, at its end *dipping towards* the centre of the mountain.

The lava beds of a Volcanic mountain may therefore contain a greater or lesser number of caverns, which serve, whenever they happen to lie below the normal line of water-level, as so many reservoirs. The mass of the lava is further riddled with fissures of all dimensions, which act as water-conduits and channels of intercommunication.

Even the more impermeable tufaceous beds contain cavities which may serve as reservoirs. These cavities, which attain dimensions of 2 feet or more, and take

---

[1] *Cosmos*, Sabine's Translation, vol. i. p. 230.
[2] *Bull. Soc. Géol. de France*, 2nd ser., vol. xxiii. p. 34.

a vertical direction, like the flues of chimneys, have been formed by the disengagement of elastic vapours during the consolidation of the beds, that consist, in the Naples district, of volcanic tuff with trachytic and other rock fragments. These decrease in importance as they trend from the central area of eruption.[1]

The dykes running in vertical lines through volcanic mountains form another structural feature bearing upon the question under consideration, for they traverse radially the beds of ashes, scoriæ, tufa, and lava wrapping round the central duct, with which they serve to establish intercommunication. Besides these great radial dykes, there is a network of small fissures or dykes branching off from them in all directions (see Fig. 2).

During the eruption of Etna in 1865, a rent was formed at the crater of Frumento, which extended in a direction away from the central cone for a distance of $1\frac{1}{2}$ mile. Scrope says that in nearly every lateral eruption of Etna, the production of such a fissure has been observed. Similar instances are not wanting in Vesuvius. In 1738 a fissure crossed the whole island of Lancerote; while in the great eruption of Hecla of 1783 the fissure which was then formed was supposed to extend not less than 100 miles in length.[2]

Scrope further remarks that "the rents thus produced in the frame of a volcanic mountain are sometimes of such a size as to cleave its whole mass in two. This occurred in the volcano of Inachian, one

---

[1] Dufrenoy, *Ann. des Mines*, 3rd ser., vol. xi. pp. 113, 120 (1837).
[2] *Volcanos*, 2nd edit., 1862, pp. 161—163.

of the Moluccas, in 1646. The crater of the Soufrière of Montserrat, and the volcanic cone of Guadaloupe both appear to have been thus split through. So also the Montagne Pélée of Martinique." Piazzi Smyth states that the cinder beds surrounding the summit of Teneriffe are traversed by dykes proceeding in radial lines from the Peak.[1] Phillips describes some of the dykes and fissures radiating from the central core of Vesuvius.[2]

These dykes have generally become fissured in cooling, and the interstices thus formed act as water conduits, which serve as so many channels to carry the water from the separate water-reservoirs in the mountain, and drain them into the central duct whenever the normal hydro-geological conditions are disturbed. At such times the dykes there contribute greatly to the discharge of water into the interior of the volcano.

Very little is known of the substrata of a Volcanic mountain. We know that Vesuvius, Etna, and Hecla stand on Tertiary strata—that some volcanoes in America stand on Cretaceous or Jurassic strata, and others on the older rocks, but of the stratigraphical details underground we have very scanty information. The only instances that I am acquainted with are the sections obtained in the boring of the two artesian wells at Naples by MM. Degousée and Laurent, of Paris, in 1865-66. These supply important data not only respecting the volcanic beds, but also respecting the sedimentary strata beneath. One well

---

[1] *Teneriffe*, p. 80.
[2] *Vesuvius*, pp. 132 and 191, and Pl. VI.

is situated in the Piazza Villa-Reale, and was carried to a depth of 1106 feet, and the other, in the gardens of the Royal Palace at Naples, 72 feet above the sea-level, was carried to the depth of 1524 feet. The details given of this latter by M. Laurent are as follows :—[1]

Section of Artesian Well in the Palace Gardens, Naples.

		Thickness. mètres.	Depth. mètres.
	1. Soil and made ground	16·50	16·50
	2. Yellow volcanic tuff	52·50	69·00
	3. Green ,, ,,	33·00	102·30
Volcanic ejectamenta.	4. Volcanic ashes, in places argillaceous, and containing numerous pebbles of trachyte .... 1st *spring*	103·40	205·40
	5. Greenish volcanic tuff	7·00	212·40
	6. Gray clay	8·10	220·50
	7. Gray marly tuff with trachyte	4·00	224·50
	8. Sandy marl with veins of lignite	25·00	249·50
	9. Gray marly and bituminous sands, with mica. 2nd *spring*	27·00	276·50
	10. Hard sandstone	1·80	278·40
Sub-Apennine strata.	11. Compact shelly marl	44·80	323·20
	12. Alternating micaceous sands, soft sandstones and carbonaceous marl ...... 3rd *spring*	48·70	371·90
	13. Micaceous marl and siliceous limestone	7·81	379·70
	14. Compact micaceous marl	53·19	432·90
	15. Compact marl with layers of limestone	25·72	458·62
Eocene strata.	16. Argillaceous limestone	2·00	460·62
	17. Macigno (a hard calcareous sandstone) .... 4th *spring*	4·08	464·70

[1] *Guide du Sondeur*, 2nd edit. 1861; vol. i. p. 137; ii. p. 496; and Pl. L.

No ascending spring was met with till the bed of volcanic ashes No. 4 was reached at a depth of 368 feet. Another spring was found at a depth of 830 feet in the Tertiary sands No. 9; another in the micaceous sands No. 12 at 1106 feet, but no spring of the desired volume was struck until a sandy bed under the Macigno was reached at a depth of 1524 feet. The discharge of water from this bed amounted to nearly 2 cubic mètres per minute, and rose about 30 feet above the surface, so that the water could be used as a natural fountain in the Palace gardens. In the Piazza the artesian waters formed another natural fountain rising 8 feet above the surface.

These wells, therefore, show the existence of one artesian spring in a stratum of volcanic *débris*, and of three similar springs in the sedimentary strata beneath. But this only gives the more powerful ascending springs; bodies of water of lesser volume, or which do not rise to the surface, escape notice in wells of this description. As the water had sufficient ascensional force to rise several feet above the ground, it must necessarily stand at its outcrop in the central volcanic area considerably higher than at the point of overflow.

In this way all the permeable beds of a volcanic mountain may be charged with water, the level of the water rising with the distance from the point of escape and with the height of the ground. Everywhere beneath the level of saturation the surface-waters will eventually fill all fissures and interstices, and lodge in them permanently, unless disturbed or drawn off by artificial or natural means (see Fig. 1, Pl. 13).

The height of the arch formed by the water-level depends mainly on the texture of the rock, or the friction to which the water is subjected. In the Chalk hills of the south of England, which are composed of a comparatively homogeneous rock, the rise of the line of water-level varies from 13 to 150 feet in the mile. In some strata it is considerably more. In the case of the irregular and complex beds forming a Volcanic mountain, the height of the water-level is subject to too many conditions to be determined accurately, except by observation, and for this few opportunities present themselves. There is, however, an available natural datum line, namely, that furnished by the escape of springs, at certain high levels on the mountain slopes.

As springs issuing from a body of homogeneous strata or from beds which intercommunicate, depend for their permanence upon the water stored up in the interior of the mountain, at a level above that of the point of escape, it follows that if there is a point of permanent escape, we may conclude that, in the ground behind, all the strata below that level are under the line of permanent saturation, and therefore charged with water; and further, this line of permanent saturation must stand the higher the further it goes into the body of mountain.

On Etna (Fig. 1), Wattershausen[1] describes a spring in the valley of St. Giacomo, near Zafarana, which he says is the only point at so high a level at which

---

[1] *Atlas de l'Etna*, Part I. and Pl. V.

a tolerably strong spring constantly issues. It at once forms a small waterfall, and runs some distance until lost in some volcanic sands lower down the slope of the mountain. The strata, whence the spring issues, are composed of alternating layers of tuff and compact lava. Its escape is due to the circumstance that there is here a ravine which cuts through the bed, and thus taps the subterranean waters. Wattershausen does not give the height of the ground, but from a section of Abich's, which passes near the spot, I infer it to be about 2000 feet above sea-level.[1] On the other side of Etna the river Simeto, or one of its tributaries, rises near Bronte, at a height of about 2200 feet above sea-level. This likewise indicates the existence of a perennial spring. I find also that at other points around the mountain at and about this level, streams commence which point to a like line of water-level.

These figures are only roughly approximate, but they constitute our only available data, and in taking a mean level of 2000 feet, I believe I am rather below than above the mark. If, therefore, we take Wattershausen's section of Etna, supplemented by Abich's, which follows nearly the same line, and gives, moreover, the height of the several points, the following diagram would represent generally the *massif* of the mountain and the position of the line of permanent saturation (*l*).

From the points of permanent issue at Zafarana and near Bronte, the line of water-level must rise in proportion as the ground rises. It is therefore

[1] Vesuvius and Etna, 1837. Sections of Etna.

possible that in the centre of the mountain the line of saturation, l, may occasionally rise higher than the Val del Bove or up to the dotted line l' (Fig. 1). The remarkable flood which swept down that great depression during the eruption of the year 1755,[1] seems explicable on the supposition of a high central water-level more readily than on the assumption of the sudden melting of a mass of ice in the interior of the mountain. For although ice may be formed and retained under a covering of lava near the summit of Etna in the manner described by Lyell, the cold would not penetrate to a

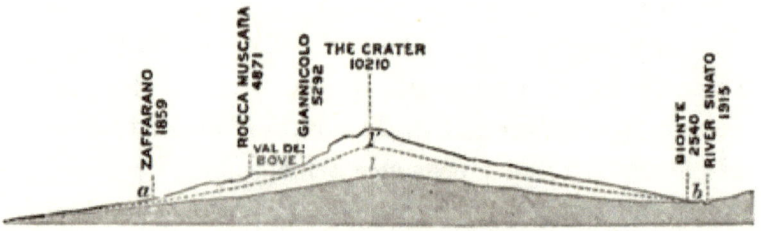

Fig. 1.—Diagram section of Etna.
The shaded part shows the extent of the mean permanent saturation.

sufficient depth to allow of the accumulation of a mass of ice of the dimensions required for so great a flood. Any body of ice would be superficial, as the increase of heat with the depth from the surface would be a bar to its existence in the interior of the mountain, independently of the heat diffused from the central duct. Nor would the sudden melting of the snow, which never lies very deep, explain the transient great and exceptional outburst of flood-waters described by contemporary writers.[2]

[1] Lyell, *Phil. Trans.*, vol. 148 (1858), p. 68.
[2] Canon Recupero, who reported on the catastrophe, came to

On the other hand, heavy rains, or a prolonged repose of the volcano, may have resulted in the exceptional rise of the level of the underground water-line to l'. If, then, one of these radial fissures, so frequently formed during the eruptions of Etna, suddenly opened in a direction to traverse the water-logged strata, the effect would be to tap and drain at once the whole of this subterranean reservoir lying above the level of the point of escape in the Val de Bove—which is situated 5,292 feet above the sea-level,—while the water, coming as it would from the very centre of the mountain, would also account for its recorded heat.

With respect to the condition of the sedimentary strata under a volcanic mountain, very little is known. M. Constant Prevost, after visiting most of the volcanic districts in Europe, concluded that they were not in general much disturbed and that the volcanoes were on lines of fissures but not on lines showing much lateral compression, or on anticlinals. With respect to Vesuvius, we should imagine that the sedimentary strata pass under the mountain and crop out in the adjacent sea-bed with little interference. The tertiary strata under Naples come to the surface on the hills further inland, and dip continuously seaward. The fact of the overflow of water in the artesian wells to the height it actually attained shows that the continuity of the water-bearing beds cannot be materially interrupted.

the conclusion that the water was vomited forth by the crater itself, and was driven out from some reservoir in the interior of Etna (**Lyell**, *op. cit.*).

The well-sections at Naples have proved that there are at least three water-bearing beds in the tertiary strata underlying the volcanic *débris*. Of these the deeper one (1524 feet) discharged at first about 600,000 gallons daily, rising 102 feet above the sea-level. So great a pressure would show that this stratum had its outcrop inland at a considerable elevation above Naples, and the volume of the spring would prove that the head of underground water above the line of sea-level was large. Consequently, as the plane of this water-bearing bed must be traversed under Vesuvius by the volcanic duct, the sides of that duct or fissure will at that point be subject to considerable hydrostatic pressure—a pressure possibly equal to that of about fifty-three atmospheres.

With respect to the ordinary mode of escape of the underground waters, the portion held in the more superficial volcanic beds escapes as springs on the slopes or at the foot of the mountain; but when they, or the underlying sedimentary strata, crop out in the bed of the sea—the surplus waters (or those annually added to the underground stores by the rainfall) will escape—after supplying the surface springs—by springs issuing in the sea-bed. When the water-passages are contracted, as in sand and sandstone, these submarine springs will be small and slow: but when large and more open, as in limestones, the discharge in the sea-bed will, as on the coast of Spezzia and elsewhere on the Mediterranean, form large and powerful springs of fresh water rising through the waters of the sea.

Such are the hydro-geological conditions of a volcanic mountain in a state of rest. The effects—when that equilibrium is destroyed can now be considered.

§ 7. Condition of the Underground Waters during an Eruption

So long as the volcano remains in a state of rest, so long will the hydro-geological conditions described in the last chapter continue unchanged. The level of the underground waters may rise; but no result is produced except that due to the trickling of some water on the yet hot surface of the plug, giving rise to the small columns of steam common during the periods of rest. After a prolonged period of repose, even these minor effects cease. It may even happen, if the crater is very deep and there has been a long interval since the last explosion, that the water-level may mount into the crater and give rise to a lake. Or it may be that, owing to the decomposition of the lava, it has formed a bed retentive of the rainfall. Such bodies of water are important in the event of any fresh eruption.

Crater lakes are not so common or of so large size in Europe as in Central America. The lake of Atitlas, 1558 mètres above the level of the Pacific, is 20 × 15 kilomètres large, and of a depth yet unascertained. The crater lake of Masaya in Nicaragua is 8 kilomètres across, and 150 mètres deep in the centre. In 1852 this lake suddenly appeared as though boiling, and a violent explosion shortly followed. The volcano of San Salvador is 2300 mètres above the sea-level, and at the bottom of its crater, which is 700 to 800 mètres in diameter, and 400 to 500 mètres in depth, is a large lake. Another large crater lake, 12 kilomètres west of San Salvador, rises level with the ground and is 200 mètres deep.[1]

The crater lakes, however, are an exceptional feature. The great body of the rainfall becomes stored within the mountain itself, and makes itself visible only by springs on the lower slopes of the mountain. As there may be a number of independent water zones, one of these may be affected by the disturbances accompanying an eruption, while others adjacent do not suffer, or suffer but little. When, however, these several zones are traversed by communicating rents the whole mass forms a common reservoir. The following section shows the irregularity of the ash beds on the flanks of a volcano, with the intercalation of lava streams and dykes.

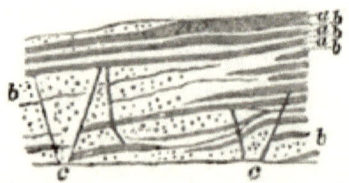

Fig. 2.—Section on the slopes of the old Volcano of Santorin (Fouqué). *a*, Lava flows; *b*, Permeable scoriæ and ashes; *c*, Dykes.

Whenever an eruption takes place the hydrostatic conditions of the volcano are disturbed, and a special class of phenomena ensue. As the upward pressure

[1] *Voyage Géologique dans la République de Guatemala et de Salvador*, par MM. Dollfus et de Mont-Serrat, 1868, pp. 103, 106, 318-20, 374-5.

of the molten lava beneath begins to exert itself, and the more solid plug yields, small cracks and fissures are gradually formed in it, into which the water lodged in the beds around, or may be in the crater, finds its way, causing an increase in the discharge of vapour, and giving rise to detonations, small at first, but increasing gradually as the heated lava breaks through the plug and comes more into contact with the interstitial waters. When the tension has reached a certain point the fissures become enlarged and through these the stored up water will be precipitated on to the hot lava in the duct, in quantity sufficient to produce those powerful and paroxysmal explosions with which the eruptions usually commence after lengthened periods of rest.

As these detonations and explosions shake the fabric of the mountain—the underground waters suffer disturbance, old water-channels are blocked up, new channels are formed—and the waters are dislodged and driven to seek new lodgments, while the water lodged in the centre of the mountain flows into the volcanic duct, flashes into steam, which is driven out in continuous explosions. The water stores immediately surrounding the central duct and crater are thus gradually exhausted, and their level is more or less lowered. As this loss proceeds, the water lodged in more distant parts of the volcanic mountain flows in to supply the void, and the explosions will be violent and prolonged, according to the available volume of water present in the mass of the mountain. As the central water stores are used up, and the level of the underground water gradually lowered, an influx from the circumference of the mountain takes place to replace the loss caused by the explosions, and this will continue so long as the body of water in the centre of the mountain is kept by the eruption lower than that in the strata and fissures at a distance. These different hydraulic conditions are represented by the diagrams —sections 1, 2, 3, Plate XIII.

In these diagrams v represents the mass of volcanic materials; p the permeable Sedimentary strata; m the impermeable strata; a, b, c the normal level when undisturbed, and a, b', c, when disturbed, of the underground waters in the volcanic mass; and b' their level in the Sedimentary strata. The arrows show the direction in which the waters flow under the normal, and under the altered, conditions caused by the eruption. In section No. 1 the points a and c are fixed levels, while b fluctuates according to the amount of rainfall and to the length of the interval between the several eruptions. Its height above a and c depends upon the distance from those points and on the resistance of the materials through which the water percolates. This of course is apart from any exceptional interfering causes. The flow of water in section 1 will, under normal conditions of repose, be outwards, where it will escape as springs at the lowest levels; and if there are no lower levels inland, the whole flow will be seaward.

As the level of b is lowered by the discharge of water into the duct of the crater, whence it is dispersed by the explosions, the underground waters will first cease to escape at a and c, and then the

DIAGRAM-SECTION OF A VOLCANO,
BEFORE (1), DURING (2), AND AFTER (3), ERUPTION

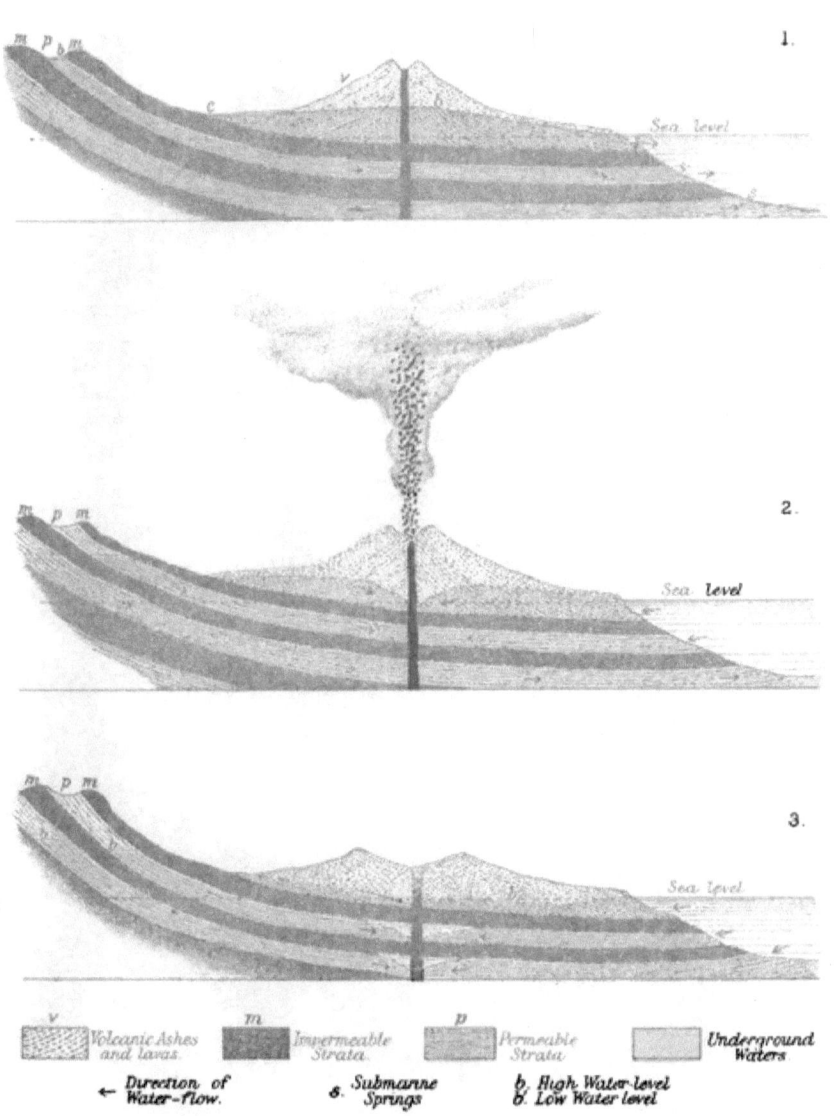

water in the centre of the mountain will set in with an inward flow towards the central duct (section 2); giving rise to a depression in the water-level in the form of an inverted cone, like that which is produced by excessive pumping in artesian wells.[1]

This flow of water towards the volcanic centre is greatly facilitated by the dykes radiating from that centre, and which, intersecting at right angles the several water-bearing masses, serve as conduits to carry the water into the main duct of the volcano. Nevertheless, there may be some areas so isolated by local conditions as to escape drainage. If, however, a large fissure were formed through the head of water ab or cb, and it happens to open out on the slope of the mountain at a level lower than the water level in the interior of the mountain, then the escape of water would be outwards, and its volume would be in proportion to the mass of the mountain above that level, or to the height of the water-level at b at the time. Such a fissure, probably during an exceptionally high water-level, might account, as before mentioned, for the flood in the Val de Bove on the slopes of Etna in 1665, and it is possible that the same cause may have produced some of the great water and mud discharges recorded of other volcanoes.

The progress of the eruption by which the water

[1] The vapour of water constitutes by far the largest part of the elastic fluids given off during eruptions, probably $\frac{950}{1000}$ or even $\frac{999}{1000}$ of the whole. M. Fouqué estimates that the quantity of vapour projected from Etna in the eruption of 1865 amounted to the large quantity of 22,000 cubic mètres, or about five million gallons daily.

level, b, is gradually lowered to b', finally determines the whole of the available drainage of the mountain into the central cavity or duct, especially if aided by radial dykes. When the dykes or fissures pass out to sea, they serve to carry the salt waters of the sea to the duct of the volcano.[1] Professor Moseley mentions[2] an eruption, that in 1877 took place on the Hawaiian coast, during which a fissure opened in the bed of the sea, in a depth of from 150 to 400 feet of water and 50 miles from Mauna Loa. This fissure was traced inland from the shore for nearly 3 miles, varying in width from a few inches to 3 feet. "In some places the sea-water was seen pouring down the opening into the abyss below."

Similarly, should the water-level b' in and under the mountain fall below both the sea-level and below the general level of the inland underground water; then not only will the springs at the surface fail and the wells run short or dry, but the *outward and seaward current will also be reversed*, and the water will flow in from the sea to the seat of the volcanic disturbance, *through the same channels* as those by which the inland waters before escaped seaward, as shown in sections 2 and 3. With the fall of the underground water-levels, the available supply of fresh water becomes gradually exhausted or the channels of communication impeded, and this con-

---

[1] Just as wells adjacent to the coast and deeper than the sea-level are subject to an influx of sea water if the pumping is carried too far, or the level of the springs too much lowered. (See *Geology*, vol. i., p. 164.)

[2] *Notes by a Naturalist on the "Challenger,"* p. 503.

tinues until, with the cessation of the extravasation of the lava, the eruption comes to an end.

When the volcano is on low ground and near the sea, and the fissures are of large size, the vast volumes of water which they admit and which flashes instantaneously into steam give rise to those stupendous explosive eruptions such as occurred at Conseguina and Krakatoa,[1] and were attended with such terrific results. It is impossible to ascribe these solely to occluded vapours. If the surface waters can produce results of that magnitude, what need is there therefore to appeal to other agencies than the surface waters in explanation of minor effects.

The access of sea-water to the interior of a volcano has long been a well-recognised fact, but whereas on Scrope's hypothesis it is supposed to find its way to the volcanic foci at great depths, and there act as the motive force in propelling the column of lava to the surface, there is, as before shown, reason to suppose that neither the surface nor the sea-waters can penetrate to those great depths; and it is further doubtful whether such motive force would be sufficient. I look upon these violent explosive eruptions as due solely to the sudden inrush of the sea-water into the crater, and there coming into contact with the highly heated and incandescent lava, it explodes with terrific violence.

The gases generated by the explosion of average gunpowder, occupy about 300 times the volume of the original powder. Gun-cotton has an expansion

---

[1] *Krakatoa*, par M. Verbreck, 1886. *Report of the Krakatoa Committee of the Royal Society*, 1888; *Trans.* extra vol.

about three times greater than that of gunpowder. On the other hand, water in passing into the state of steam or vapour, expands to 1,700 times its original volume. This ranks more nearly with the expansive force of powerful explosives, and furnishes a force sufficient to explain the blowing up of the mountain and grinding the ejectamenta into dust without calling in the aid of occluded water.

§ 8. Transmission of the Surface Waters into the Volcanic Duct.

To return to the first stage of the eruption. The lava, as it rends and crashes through the plug, comes into contact with the water lodged in the cavities and porous strata around the duct; explosions and detonations follow, violent in proportion to the supply of water. More or less of the lava is hurled into the air and scattered as scoriæ and ashes, and the explosions continue so long as the surface waters find their way to the escaping lava, but as the supply becomes gradually exhausted, the detonations diminish in power and number, until they finally cease from lack of supply. But the extravasation of the lava often continues long after this exhaustion of the water supply, showing the independence of the two causes; for were the flow of the lava dependent on the escape of occluded vapour, the flow of lava would be accompanied to the last with explosions and detonations due to the escape of the extruding agent.

This influx of water from the surrounding beds into the volcanic duct, its sudden flashing into steam and the violence of the explosions during the early stages of the eruption, are easy to conceive; but greater difficulties attend the succeeding stages, when the column of lava has ascended higher and fills the duct, and the level of the underground water has become lowered. Water must then gain access through the walls of the duct into the fluid lava as it ascends. Of the actual underground conditions we must ever be ignorant, and experiment at present guides us but a short way. Any inquiry must therefore for the present be more or less conjectural.

At certain depths the introduction of the water into the lava must depend on other conditions than hydrostatic pressure, such, for instance, as capillarity, and the elastic force of vapour. If, however, the state of rest is disturbed, and the lava in the duct begins to move upwards, then the change of conditions from the static to the kinetic immediately destroys the former balance, and the lateral pressure of the lava in the duct being no longer equal to that of the superheated vapour in the strata, this vapour is driven into the yielding mass of molten lava and ascends with it to the surface.

As the escape of this high tension vapour relieves the pressure on the water at its back, a portion of this water is driven in to replace it, and is in its turn converted into high pressure steam. Thus successive increments of vapour are driven into the lava, causing continuous explosions.

Mallet was of opinion[1] that capillary infiltration

[1] Palmieri's *Eruption of Vesuvius in* 1872. Mallet's Introduction, p. 52.

goes on in all porous rocks at enormous depths, and that the deeply seated walls of the volcanic ducts leading to the crater may be red hot, and yet continue to pass water from every pore (like the walls of a well in chalk), which is flashed off into steam, and, unable to return by the way the water came down, escapes through the duct and crater. I doubt whether this can happen during a state of undisturbed statical pressure, but it may follow on such disturbances as those just described.[1]

Under these circumstances, it is conceivable that the surface waters may readily be carried down through the upper cooler strata down to the strata surrounding the volcanic duct; but no amount of available vapour tension could force it back through the same depth of strata against both friction and capillarity. At the same time when the elastic tension of the vapour of the water reaches the critical point, or such higher temperature that the vapour tension exceeds the hydrostatic pressure, the further progress or descent of the water will be stayed. At this point,

[1] To a certain point the influence of capillarity has been proved experimentally by M. Daubrée, who found that water placed on a disk of fine-grained (Triassic) sandstone, fastened over a vessel filled with steam under pressure of nearly two atmospheres, infiltrated into the underlying vessel against that pressure. He further noticed that in consequence of the heat the action was more rapid than it otherwise would have been; and—making the experiment inversely—he observed that the vapour placed under a pressure of several atmospheres in the lower vessel did not transude through the disk left dry on the upper surface. As before pointed out, however, capillarity is adversely affected by a rise of temperature, and is comparatively inoperative at very high temperatures.—*Géologie Expérimentale*, 1879, p. 236.

whatever it may be, while the descent of water to the *volcanic foci* through the solid crust remains an impossibility, these other causes come into operation, which render its introduction into the *volcanic duct*, even at considerable depths, possible. In any case, when an equilibrium is established between the vapour tension and the hydrostatic pressure, no change will take place unless that condition of equilibrium be disturbed.

The influx of water into the volcanic duct will therefore be materially influenced by the changes of level, or oscillation, which the column of lava undergoes during eruptions, as the variation of pressure caused by this state of changing equilibrium allows the vapour pressure to predominate at one time, while the resistance to it by the column of lava will be in excess at other times.

§ 9. The Ejection of Blocks of Rock during Eruptions.

The explosion of the pent-up steam into the lava involves, amongst other effects, disturbances of the encasing strata whenever any portion of the water or vapour explodes behind blocks of the rock forming part of the walls of the duct.

When the underlying sedimentary strata consist of limestones or sandstones traversed by vertical joints and cracks, in which water can lodge, any alteration in the pressure of the lava column in the duct due to its upward flow or its oscillation, causes the water behind the pieces of rock to flash instantaneously into steam, and explode with violence. This drives blocks of these rocks into the ascending column of lava, while the cavities they leave are at once filled with the ascending lava. As this casing of the duct remains at depths for a lengthened period exposed to the intense heat of the molten lava-column, it becomes metamorphosed, so that the blocks of rock with the injected lava casing, blown into the ascending column of lava, are found to have undergone molecular changes, such as are exhibited by the blocks of altered tertiary and cretaceous limestones, and of the old lavas that occur amongst the earlier ejectamenta of Vesuvius on the slopes of Somma, and which contain a very varied group of minerals, the result of metamorphic action. These minerals are such as could only have been found in the presence of water. This would not have been the case had the blocks been encased in the molten anhydrous lava at great depths.

Similar metamorphosed blocks are found in the old lavas of many volcanoes. Fouqué mentions that in the eruption of Santorin in 1866 there were enclosed blocks of a lava of which the composition differed considerably from that of the more recent lavas, and that they contained the minerals common in the old lavas of that island,[1] while blocks of quartz micaschist occasionally met with had undergone little or no change.[2]

Summary.—It would thus appear that a volcanic duct passing through a certain number of permeable strata charged with water, and then through volcanic

[1] Fouqué, *Santorin*, p. 12.   [2] *Ibid*, p. 10.

materials also charged with water, is surrounded at all points by, as it were, an explosive substance under repression that only requires a disturbance of the equilibrium, under which it exists when the volcano is in a state of repose, to explode with violence. Further, as the superheated vapour in the encasing strata forces its way into the ascending lava, adjacent portions of highly heated water are driven forward to replace that which has escaped—an abstraction that gradually lowers the water-level in the mountain. As the lava continues to ascend, the injection of the vapour, and the explosions of successive portions of water and steam, proceed uninterruptedly, and the volcanic duct thus becomes, as it were, the centre to a battery keeping up an incessant volley; for the supply of water at first is comparatively unlimited, so that being replaced as quickly as it is exploded, there is no cessation in the action until the source is drained or stopped, or till the lava ceases to flow. If water is lodged behind blocks or fragments of rocks or of older lava, then these blocks will be blown and driven into the outflowing lava current. This seems to be a more probable explanation than that they are mechanically torn or wrenched off the sides, although in the first stage of a new volcano there may often be great mechanical action and ejection of *débris* in clearing the passage, as in the case of the large quantity of *débris* of Devonian strata in the old volcanoes of the Eifel.

In considering the various phases of this problem, it is only too apparent that while the hydro-geological conditions admit of investigation on known principles, the thermo-dynamical conditions are involved in much obscurity, and are more hypothetical. The warrant for any hypothesis depends, however, upon whether the observed phenomena accord with the inferences that should follow on the assumptions, and for this I think sufficient cause can in this instance be shown. For the more or less deep-seated subterranean detonations and thundering that accompany most eruptions, and the paroxysmal explosions accompanied by enormous ejections of steam and ashes from the crater, are the necessary consequences of an influx of water into the volcanic duct under the conditions we have described; while a term is placed to the continuance of the eruption in the circumstance that the water supply being external and independent, whenever that supply is exhausted by expulsion through the escaping lava, the explosions must cease, although the eruption of lava may proceed for a longer period. The hypothesis agrees also with the fact that as a rule the eruptions are more paroxysmal the longer the interval of rest, for the filling of the underground reservoirs exhausted by the previous eruption is a question of time, and the greater their water-stores, the greater and more powerful the explosions. If the escape of vapour were due to steam occluded in the lava in the volcanic foci, there could be no violent explosive eruptions, for the steam coming up heated and in a state of vapour would expand gradually as the pressure was removed, and not so suddenly as when water flashes into steam, so that the eruptions would lose much of their violent character.

The apparently conflicting observations of Ehrenberg respecting fresh-water diatoms in the volcanic ejections of some islands, and of marine diatoms in others, admit of ready explanation on this hypothesis. For whenever the inland underground surface fresh-waters stand at a certain height above the sea level, so long will those waters dam back and keep out the sea-water, but whenever their level is brought below that of the sea level, then inevitably will the sea-water flow inland until the level is restored. Thus fresh-water remains accumulated during a period of repose, may be ejected during the early stages of an eruption, and may be succeeded by marine ejections when the exhaustion of the fresh-water springs leads to the influx of the sea.

The extinct volcanoes of Auvergne, the Eifel, Hungary, and Central Asia afford coroborative evidence of the need of surface waters to maintain a volcano in a state of activity, for in these areas great fresh-water lakes—like those which lately existed in the volcanic districts of New Zealand—of late tertiary age formerly existed. This connection has often been noticed. It was no doubt in consequence of these lakes gradually drying up or being drained by geological changes, that these volcanoes became extinct. Such phenomena are clearly independent of occluded water, otherwise there is no reason why the eruptions should not have continued to the present day. Now most of the great volcanoes are confined to islands and coast lines.

The effects of this proximity on these volcanoes, and of the access of sea-water is shown, as is well known, not only by the presence of chlorides and other products of its decomposition in the emanations from the lava, but also in many cases by the presence of sea salt itself. This, however, is not always the case. The great South American volcanoes seem to be dependent upon the inland waters alone, for M. Boussingault was unable to detect in the fumaroles of those volcanoes any traces of chlorides.

But if water only plays a secondary, though admittedly important part, to what are we to attribute the motive power which causes the extravasation of the lava?

§ 10. Primary Cause of Volcanic Action.

If the presence of occluded water in the volcanic foci be not the primary cause of the expulsion of the lava, to what other cause is it to be attributed? I see none but a modification of the old hypothesis, namely, that of the contraction of the solid crust of the earth upon a yielding and hot nucleus. A main objection to this hypothesis rests chiefly on the fact that if, as was at the time assumed, the whole nucleus beneath the solid crust consisted of a molten fluid, it would be subject to tides that would lessen or neutralise the surface tides. Sir William Thomson (now Lord Kelvin) and the late Mr. Hopkins have proved that the earth possesses a rigidity incompatible with a fluid nucleus. At the same time, objections have been taken by other physicists to the hypothesis of an entirely solid globe,[1] on the

---

[1] See the various papers on this subject by Hennessy, Haughton, O. Fisher, Elie de Beaumont, Delaunay, Roche, and others. Newcomb also admits the probability of a thin outer crust.

grounds, amongst others, that the question has been dealt with on the assumption of a perfect fluid, and that not sufficient allowance had been made for other causes. (For the fuller consideration of this subject see Paper No. V. p. 147.)

For the geological requirements, neither a perfectly fluid substratum nor a molten nucleus are needed. The hypothesis of a central solid nucleus is perfectly compatible with geological phenomena, provided that the solid nucleus is surrounded by a molten yielding envelope—not fluid, but viscid or plastic; nor is it necessary that it should be of any great thickness. The relative proportions of the two are questions for physicists. The late M. Roche proposed a solution based on the astronomical and physical conditions of the problem. Briefly assuming the earth to consist of a solid centre with a density of about 7·0, and of an outer layer, consisting of a fluid substratum with a solid crust, having a mean density of 3·0, he found that these two outer layers should together have a maximum thickness equal to one-sixth of the earth's radius, or 660 miles, but he left the question of the relative thickness of the two open to other considerations.

On geological grounds the solid crust need not have a thickness of more than 20 to 30 miles; while, on the same grounds, the dimensions required for the underlying molten layer, instead of being a substratum having a thickness measured—not by hundreds, but by tens of miles—would be sufficient to fulfil the necessary conditions.

It is quite possible, as suggested by Scrope, that

owing to pressure the fusion-point of lava at great depths is so much higher than at the surface, that the lava may, and possibly does, exist at depths in a highly viscid state, and only becomes fluid as it rises to the surface and the pressure is removed. A state of viscidity accords with the slow rate of movement and steadiness of the great continental elevations and depressions—changes in close relation one with another, and which may arise from the slow transference from one area to another of a partially resisting plastic medium within confined limits.

Under these conditions that portion of the molten layer, provided it were of moderate thickness, would, when locally compressed between the outer solid crust and the inner solid nucleus, expand laterally, and the mass displaced would be transferred to that part of the adjacent area where the outer crust would yield most readily to deformation and upheaval.

A compression in one part should therefore be followed by expansion in another, and by opposite deformations of the crust over conterminous areas. These effects are exhibited in the great continental upheavals and depressions so rife in the times immediately preceding our own, and still in a measure of perceptible action in many districts; as, for example, in the instance of the slow uplifting of the northern areas of the Scandinavian and Greenland peninsulas, and the subsidence of their southern areas. On the assumption above named all these movements are due to the agency of the same force, namely, that resulting from the contraction due to secular refrigeration; but while in one case it produces excessive lateral compression,[1] in the others it shows in turgid swelling of the crust.[2]

The agency of water is, I conceive, confined, therefore, to the secondary effects described—effects perfectly independent of the forces that produce the extravasation of the lava; and, while with the thinner crust of former times, there would be a more frequent extrusion of the molten rock, there are probably, with the thicker crust now formed and consequently its greater resistance, greater forces stored in the explosive eruptions of the present day.

The loss of terrestrial heat by radiation is now exceedingly small. But small as this loss is, it cannot take place without producing contraction, and Cordier long ago estimated that, supposing five volcanic eruptions to take place annually, it would take a century to eject so much lava as would shorten

---

[1] This is I am aware contested, and other reasons are assigned for mountain elevation, but they seem to me to present greater difficulties than the old contraction theory.

[2] This constitutes an essential difference between the disturbances exhibited in mountain and in volcanic areas. Both may result from the contraction due to secular refrigeration, but the one is a process of excessive lateral compression, and the other of turgid swelling of the crust. In both cases there is tension, although of a different character. In the latter it is slow and uniform in its action, and where there are points of comparatively slight resistance, as in volcanic ducts, it then readily finds relief in the expulsion of the lava, which is only prolonged until the equilibrium is restored. The extrusion of lava then ceases, and the volcano lapses into a state of rest only to be broken when again there has been further accumulation of the necessary energy.

the radius of the earth to the extent of 1 mm., or about $\frac{1}{25}$th part of an inch.

I conclude **that** the hypothesis originally propounded, namely, that volcanic phenomena are dependent on **the** effect of secular refrigeration, is, with certain modifications, the one that best meets the necessities of the problem.

V

ON THE THICKNESS AND MOBILITY OF THE EARTH'S CRUST FROM THE GEOLOGICAL STANDPOINT

Geologists and Physicists still hold from their different standpoints divergent views respecting the thickness to be assigned to the crust of the earth. Although the present stability of the earth's surface renders it evident that the hypothesis of a thin crust resting on a nucleus altogether molten and fluid is untenable, it is equally difficult to reconcile certain geological facts with the hypothesis of a globe solid throughout, or even of a very thick crust.

The phenomena of the tides have not yet been determined with sufficient accuracy to settle definitely the moot question whether the rigidity of the crust is perfect, or whether it yields to some very small extent to the deformation that might be caused by slight internal tides. There are, on the other hand, certain geological and volcanic phenomena which are not only incompatible with an entirely solid globe, but which would seem to be explicable only on the hypothesis of a thin crust and a slowly yielding plastic substratum. Whether a crust and substratum of this nature would, under certain conditions, offer sufficient resistance to produce a *quasi* rigidity such as would accord with existing physical conditions, is yet a debated problem.

The phenomena on which on geological grounds I should chiefly rely in proof of a crust of no great thickness are—

1st. *The flexibility of the crust as exhibited (a) in the uplift of Mountain Chains, and (b) in the elevation of Continental areas.*
2nd. *The rate of increase of temperature with the increase of depth from the surface.*
3rd. *The volcanic phenomena of the present day, and the welling-out of the vast sheets of trappean rocks during recent geological periods.*

(*a*.) It is important for our object to note that not only has mountain-upheaval gone on through all geological time, but that many, if not most, of the great mountain chains have been raised during the latest geological periods. As instances of these may be named—

1. The elevation of the Pyrenees, which, although commenced in Palæozoic times, attained its maximum intensity and development in Oligocene, while minor movements continued to Miocene, times.
2. The main elevations of the Rocky Mountains and portions of the Andes took place during the earlier Tertiary periods, but they were raised to their present height so late as in Miocene and Pliocene times.
3. The researches of the geological survey of India[1] show that, although the elevation of the Himalayas commenced in Palæozoic times, the special great Himalayan disturbance is of Post-Eocene age; while in the Sub-Himalayan

---

[1] Medlicott and Blanford's *Geology of India*, pp. 569, 570.

ranges there is a large amount of disturbance of Post-Pliocene date.[1]

4. The elevation of the main axes of the Alps (although, like the others, begun earlier) took place in Miocene times, and was prolonged to as late as the Post-Pliocene period, or to the time immediately preceding the Quaternary period.

It is only necessary to look at the section of any mountain chain to see the enormous amount of squeezing and crumpling the strata have invariably undergone, and the succession of folds of vast magnitude into which they have been thrown (see Fig. 5, p. 236). In the Alps there are seven, if not more, of these great folds, each constituting a mountain chain. In a straight line across they measure about 130 miles; but, if the strata were stretched out in the original planes, it is estimated that they would occupy a space of about 200 miles.

Le Conte states that the coast range of California consists of at least five anticlines, and as many synclines so closely compressed that a width of 15 to 18 miles of horizontal strata has been reduced to 6 miles.

These are common geological facts. I need add but one more instance on account of the magnitude of its scale.

The Hon. Clarence King, speaking of the plication of those parts of the Rocky Mountains which lie in Wahsatch and Uinta, estimates that the folds there measure 40,000 feet from summit to base.[2] What must have been the contraction in horizontal distance

---

[1] It is a question even whether the earth-movement along this great axis of elevation has yet wholly ceased.

[2] *Geology of the 40th Parallel*, vol. i. p. 761.

where the strata form not one but several folds, the crown of whose arches attain a height to be measured by miles!

It is difficult to see how these corrugations of the earth's crust are to be accounted for, unless we assume that the crust rests on a yielding substratum, and that is of no great thickness. For if the earth were solid throughout, the tangential pressure would result not in distorting or crumpling, but in crushing and breaking. No such results are to be seen, and the strata have, down to the time of the youngest mountains, yielded, as only a free surface-plate could, to the deformation caused by lateral pressure. Freedom and independence of motion are evident in these wonderful contortions and inversions of the strata, and for that result a soft and yielding bed on which the crust could move as a separate body is necessary. The evidence that such a yielding plastic bed does exist is shown in the fact that the central axes of mountain chains are so frequently composed of crystalline rocks, which were evidently originally in a viscid state; while the injection into the overlying sedimentary strata of dykes and veins from those rocks, indicates still more clearly the fluidity of the fundamental base.

These facts are so patent to Geologists that it may seem almost superfluous to adduce them. Let us suppose not entire solidity, but a crust 800 miles, or even half 800 miles, thick. What would be the magnitude of a mountain chain resulting from the crumpling and upthrow of such a mass of rocks? Where have we evidence in the latest of our mountain chains of the existence of such masses? Nowhere do the disturbed and tilted strata point to a mass more than a few miles thick. If the crust had the more excessive thickness suggested by Physicists and by some Geologists, we should have had mountains if not of greater height, at all events of greater breadth; for if a solid plate of any kind be broken and the fractured edges turned up by reciprocal pressure in presence of a resisting body beneath, the width of the protruding mass will bear a definite relation to the thickness of the plate. If, on the other hand, the plate is sufficiently pliable to yield without fracture and should be bent into folds, the height of the arches and the width across the folds will in like manner be proportionate to the thickness of the plate. Where have we evidence, even in the most recent of mountain chains, when the earth was approaching its present conditions of rigidity, of a shell of 700, or 500, or 100, or even of 50 miles thick? Would it not rather appear that a crust even of 30 miles is in excess of what the height and breadth of any mountain chain would, on this finding, indicate?

To the first part of the argument, it may be rejoined that the existence of a thinner crust and of a fluid nucleus during geological times is not contested, and that the conditions of solidity and rigidity are only applicable to the globe as it at present exists. But the observation loses its point when we consider that the cooling must have been slow and gradual throughout all time—that the formation of mountain chains has been intermittent with long intervals of repose—that the last-formed chains show no change in the character or diminution in the intensity of the forces to which they are due. There is, in fact, nothing to indicate such a sudden accession of solidity as would be involved by the assumption of the free play of the crust during Tertiary times and its entire rigidity now. Besides certain forms of the forces co-ordinate with a cooling globe still continue in visible action. If one form of a force dependent on a common cause remains in operation, we are scarcely justified in assuming that another consequence arising from the same cause, though dormant, is extinct. Our limited experience suffices to make us acquainted with the more persistent effects, but altogether fails to compass those which are intermittent, with intervals of uncertain duration.

(*b.*) With respect to those widespread movements which result in great superficial or continental upheavals—movements which, of frequent occurrence in all geological times, have not altogether ceased in our own times,—it will be sufficient if, for our purpose, we go back to those which took place towards the close of the Tertiary period.

We have only to look at a geological map of the world to see that a very large portion of the existing continents have been under the sea during early Tertiary periods. South-eastern England, a large portion of France, great part of Spain and Italy, and much of Central Europe, have, since the Eocene period, undergone movements of elevation, *en masse* with little disturbance of the strata from their horizontal position over large areas, and these movements have been prolonged down to Miocene and Pliocene times. Great parts of Asia, Australia, and of the seaboard of North and South America, have also been under the sea and raised at different times during various Tertiary periods.

But our more immediate object is to show that the flexibility of the crust has been continued without break and over large areas down to the latest times, and that the older changes link on to changes in progress in our own period.

The presence of shells of recent species at certain elevations in Central England and Wales tends to prove that those areas have undergone an elevation of not less than 1,200 to 1,500 feet since the inset of the Glacial period. Ireland and Scotland have undergone similar changes in the same period; and so likewise has the North American continent. The rock of Gibraltar has been raised 600 to 900 feet at a late period. The *massif* of Scandinavia, the long coast-lines of the Pacific side of South America, have been raised 500 or 600 feet or more during the lifetime of existing species of mollusca. Other portions of the Pacific basin islands have, equally late, been raised 200 to 800 feet or more, and the coasts of Arctic America and Asia exhibit conclusive evidence of similar recent elevation. These movements bring us down to the threshold of the present times.

Some change of level has been ascribed to the rise of temperature at depths, caused by the shifting upwards of the underground isotherms, in proportion as the strata have increased in thickness by the successive addition of fresh sediment. This, as suggested by Babbage and Herschel, may be a true cause under certain conditions of thick accumulation of strata, but must necessarily, as the expansion of rocks by heat is so small, be limited to very moderate vertical displacements, probably not exceeding 50 to 100 feet, if so much.

Alterations of level arising from this cause are also not likely to have taken place in Glacial or Pleistocene times, as the sedimentary matter then deposited was generally limited to irregular masses of sand, gravel, and shingle, or to a few raised beaches and shell-beds only a few feet in thickness; while the submarine beds rarely reached a thickness of more than 100 or 600 feet.

The Effects of Conductivity on Temperature

This raises another question in connection with the increase of depth and of temperature. With respect to the latter, there is a general agreement as to the fact, but a difference of opinion as to the rate of increase and as to the value of the observations, which have given a rate of increase varying in round numbers from 30 to 100 feet for each degree F., or a mean taken by some of 50 and by others of 60 feet.

The variation in the results depends not only upon the many disturbing causes before described, but also on the varying conductivity of the strata. While all the observations go to prove the general fact of an increase of temperature with depth, very few have included this cause as one of the factors in the case. Notwithstanding the care taken by some of the earlier and all the later observers against these disturbing causes, they are so numerous and often so difficult, if not impossible, to guard against, that only a limited number of observations can be relied upon to give the more exact data required. I have gone elsewhere[1] into the various considerations affecting this question, so that they need not be discussed here. The conclusion at which I have arrived is, that the observations best to be relied on show an increase of about 1° F. for every 48 feet of depth, though there is reason to believe that in some cases it may be even more rapid than this.

Whether or not the rate of increase diminishes or increases with the depth is a question which requires further investigation. The few observations that have been made throw little light on the subject. They certainly do not tend to prove that there is any diminution in the rate of increase of temperature. There are some reasons, on the contrary, which would lead to an opposite opinion.

The conductivity of Chalk, Sand, Clay, and some Sandstones, through which so many of the bore-holes and shafts in which the observations have been made have been carried, is low. It has been inferred in consequence that the isothermal lines of depth are closer together in these rocks than they would be in the underlying crystalline and igneous rocks, and therefore that the sedimentary rocks give us a more rapid rate of increase than would be found to obtain in the older and igneous rocks at greater depths. But it must be remembered that in general the experi-

[1] Paper No. VI.

ments on the conductivity of rocks have been made with specimens carefully dried, whereas the underground rocks are charged down to considerable depths with water. The experiments of Messrs. Herschel and Lebour (*postea*, p. 240) show that there is a remarkable difference between the conductivity of rocks in their dry and wet state. Thus, taking the rocks just named, these differences are as under, k being the thermal conductivity, and r the thermal resistance.

	Dry.		Wet.	
	k.	r.	k.	r.
Quartzose Sand	0·00105	952	0·00820	122
Clay	0·00250	400	0·00370	270
New Red Sandstone	0·00250	400	0·00600	166 ?

It appears by this that sand, which is one of the worst conductors when dry, becomes one of the best when wet, being only exceeded by rock salt, vein quartz, and quartzite; while the mean conductivity of the three substances above named exceeds when wet ($k = 0·00597$, $r = 186$) that of many crystalline rocks,—the average of five varieties of granite being $k = 0·00584$ and $r = 173$. With several Palæozoic limestones k averages 0·00572 and r, 177.

Experiments on wet Chalk, Oolite, and Coal-measure sandstones are wanting, but as the water of imbibition of these several rocks stands as under, it is to be inferred that when saturated the conductivity of three of them at least would be raised in proportion with that of the above—

	Per cent. of water.		Per cent.
Coal shale	2·85	Chalk	24·10
Sandstone	4·37 to 13·15	Cal. freestone.	16·25

On the other hand, the harder crystalline, calcareous, and igneous rocks imbibe but little water, as the following list shows:—

Granite	0·06 to 0·12	per cent.
Devonian Limestone	0·08	,,
Basalt	0·33	,,
Slate	0·19	,,
Quartzite	0·66	,,

Then, again, the conductivity, even when dry, of the "ganister" and other hard Coal-measure sandstones is found to be higher than that of either the metamorphic or igneous rocks, averaging, taking four varieties, $k = 0·00737$, $r = 1·36$.

It seems, therefore, probable that the unaltered Sedimentary strata must often, under their normal underground conditions, be as good, if not better, conductors of heat than the Crystalline and Igneous rocks, and, if so, the thermometric gradient would rather tend to become more rapid in passing from the former into the latter than otherwise.

But there is another cause which, when the subject comes to be more fully investigated, will have to be considered—that is, the change of conductivity produced by heat. Forbes found that the conductivity of iron varied with the temperature as under:—

0° C.	0·0133
100° ,,	0·0107
200° ,,	0·0082

This shows a percentage decrease in the conductivity of iron in passing from 0° to 100° C., of 24·5, which agrees nearly, according to Professor Tait, with the empirical law that the conductivity is inversely as the temperature. Looking at other physical properties which the metals and rocks have in common, it is not improbable that the latter may have also this other property, although, perhaps, in a modified form. It is the more probable, inasmuch as the proportion of iron present in the igneous rocks (commonly 10 to 15 per cent.) is larger in those which are supposed to be the more deeply seated; and considering how largely the density of the mass of the earth is in excess of the density of the crust, there is reason to believe that the proportion of the metals increases with the depth. These conditions must materially affect the thermometric gradient.

After eliminating those observations of which the results are affected by these various causes of interference, it would appear that the sedimentary rocks possibly do not possess a lesser power of conduction than the igneous and crystalline rocks which underlie them, so that the rate of increase of temperature need not on this account be less rapid.[1]

Assuming a uniform rate of 1° F. for every 48 or 50 feet of depth, the heat at a depth of about 30 miles would be such as to fuse the basic rocks, and this has frequently been taken as a measure of the

---

[1] This is also the conclusion which may be drawn from the deep (4,172 English feet) and remarkable boring of Sperenberg, which first passed through nearly 300 feet of gypsum, and was then carried entirely through rock salt of which the conductivity is high (0·0128). Nevertheless, the rate of increase was found to be 51·5 feet per degree F. Had the rock been of lesser conductivity the rate of increase could hardly be otherwise than more rapid. See Brit. Assoc. Reports for 1876.

probable thickness of the Earth's crust. But there is the uncertainty just named whether the rate may not increase more rapidly, while other phenomena indicate that this even is too great a thickness to be in accordance with observed geological facts.

May it not also be a question whether, as I have before suggested, the intense cold of the Glacial period has not so affected the outer layers of portions of the earth's crust that to a certain depth the rate of cooling is now abnormally slow, owing to the excessive refrigeration the crust then underwent. If that be the case, might not the rate of increase of temperature at greater depths be more rapid than that which other observations have led us to assume, and the thickness of the crust even less than that above named? Such a conclusion would be more in harmony with geological phenomena, as we shall proceed to show.

Volcanic Action incompatible with a Thick Crust

A third objection to a thick crust is the difficulty of reconciling such a condition with the effects of volcanic action. This question hinges upon the phenomena connected—1st, with recent volcanoes; 2ndly, with the great outwellings of trappean rocks during the later Tertiary period; and, 3rdly, with the character and extent of the changes of level in the areas so affected.

1. The first point relates especially to the difficulty of imagining that a column of lava could traverse a crust 800 to 1,000 miles thick without the loss of so

much heat as to cause the lava to lose its fluidity and consolidate before it could reach the surface.

To meet this difficulty Mr. Hopkins suggested that the solid crust contained at various depths beneath the surface cavities filled with fluid incandescent matter, either entirely insulated or perhaps communicating in some cases by obstructed channels, and that in these subterranean molten lakes the volcanic foci originate.[1]

To this it may be objected that differences in depth from the surface and the existence of separate molten lakes is scarcely compatible with the singular uniformity, as a rule, of the volcanic rocks over the whole globe; as little or no pressure could be exercised by such a crust, the only available force for the extrusion of the lava would be the occluded vapour of Mr. Scrope, the objections to which hypothesis I have already named.

2. The enormous outwellings of trappean and volcanic rocks which took place at intervals during the Tertiary periods and continued down to Quaternary times, indicate the existence of a fluid magma underlying the solid crust coextensive not only with the existing volcanic areas, but also with those older ones. This would spread the volcanic phenomena over tracts so large and so numerous that the isolation of molten lakes as separate and independent local igneous centres would seem very improbable.[2]

---

[1] *Op. cit.* p. 54.

[2] See also the Address of Sir Wm. Thomson (now Lord Kelvin) in Section A, Brit. Assoc., 1876, in which the evidence regarding the physical condition of the earth is reviewed.

In this country the great basaltic plateau of the North of Ireland is 600 to 800 feet thick, and extends over an area of about 1000 square miles; those of Western Scotland are of about the same extent, and it is certain that both had originally a much wider range. In Central France there is another large basaltic area of yet more recent date. There are others of great extent in Hungary and Central Italy. They cover also large tracts in Asia Minor, Africa, New Zealand, Australia, and America. But to confine ourselves to two instances on a grand scale we may take the great plateaux of Central India and of North-west America.

In India these plateaux stretch for a distance of 500 to 600 miles from north to south, and 300 to 400 miles from east to west, covering, according to the reports of the Indian Survey, an enormous area of not less than 200,000 square miles.[1] They have a general thickness of from 2000 to 3000 feet, and it is estimated that the total thickness of all the beds amounts to not less than 7000 feet. They are of late Cretaceous or early Eocene date, and consist of a succession of beds spreading probably over a long period of time.

In North America vast sheets of basaltic rocks form the high plateau of Utah, while on the Pacific slopes immense regions have been flooded by outpourings from fissures at successive times from the close of the Miocene down to the Quaternary period. In Columbia these basaltic rocks have a thickness of from 1000 to 3000 feet. In parts of Colorado they are 4000 feet thick, and stretch over a tract some 700 to 800 miles

[1] Medlicott and Blanford's *Manual of the Geology of India*.

in length by 80 to 150 miles in width, covering 120,000 to 150,000 square miles of surface.

Extensive as are the ejections of some volcanoes at the present day, and great as are some of the individual lava streams, the sum total is small compared with any one of these older extrusions. Lyell instances as amongst the most remarkable of the modern lava streams that formed during the eruption of Skaptán Jokul, in Iceland, in 1783. He states that it formed two branches of the relative lengths of 50 and 45 miles, and of the extreme breadths of from 12 to 15 miles. Taking the mean breadth, we have an area of about 500 square miles with an average thickness of about 100 feet, and an occasional one, where it filled gorges, of 600 feet.

We have no means of judging of the separate lava-flows of geological times; we have therefore to take the total areas covered by the old basaltic outpourings, and compare them with recent volcanic eruptions. The two most extensive modern volcanic districts are those of Hawaii and Iceland. The area of Hawaii is about 3800 square miles and is entirely volcanic, and that of Iceland is 37,800 square miles, of which the volcanic beds occupy about one-third to one-half. At the outside, therefore, these modern eruptions are spread over no area larger than 20,000 square miles, or an area only equal to one-tenth and one-seventh of the old Indian and American basaltic areas.

If these vast erupted masses had been derived from local molten lakes, the extravasation would have caused a diminution in their masses which, as the loss could not be made good by drafts from adjoining areas must necessarily have led to a caving in and subsidence of the crust above these lakes to an extent proportionate to the volume of the extravasated matter. But so far from this being the case, the areas of these great basaltic outdwellings are almost always areas of elevation. The basaltic area of the Deccan forms vast plateaux which attain a height of between 4000 to 5000 feet, and although the intercalated sedimentary strata are mostly of land and freshwater origin, there is reason to believe from the circumstances that on the borders of the same district some of the associated beds contain estuarine remains, that the area was, immediately prior to the eruption, much nearer the sea-level.

In America also the basaltic plateaux rise gradually to heights of from 2000, 3000, and 4000 feet, and in some cases even attain the height of 11,000 feet or more. Similar evidence, though on a smaller scale, is afforded by the basaltic plateaux of Ireland and Scotland, of Central France, and of other countries which form, relatively to the surrounding districts, more or less elevated tablelands raised above the sea-level. These areas of eruption are, therefore, areas of elevation—not as in mountain chains by lateral squeezing and an upward thrust along narrow belts, but by elevation *en masse* of wide or continental portions of the earth's crust, possibly accompanied by fracture but without, necessarily, contortion.

It follows from the fact of the discharge of the volcanic matter being attended by upheaval and not by depression, that the igneous matter ejected is not only replaced, but that it is replaced by a quantity larger than that which is lost by extravasation. This could only be effected by supplies from adjacent areas of similar matter—in other words, it indicates that there must be a common fluid or viscid substratum, yielding to depression in some areas, and to upheaval in others, the loss in the one case being counterbalanced by centralisation in others. Apart from the great movements which raised the basaltic area of the Deccan, Dr. Blanford[1] states that in the Indian Peninsula there is evidence bringing down movements of elevation to the extent of 100 to 200 feet to so late a period as the old raised beaches (of Pleistocene age). On the other hand, the Maldive, Laccadive, and Chagos groups of atolls and coral reefs in the sea to the south of India indicate a slow depression. There is also unmistakable proof of a recent sinking of the land on the Arabian coast near the mouth of the Persian Gulf.

Though attended with more uncertainty, there is reason to believe, as suggested, I still think with great probability, by Darwin, that great coral areas of the Indian and Pacific Oceans have long been areas of subsidence,[2] while adjoining volcanic areas have been areas of elevation. In several cases areas, once areas of depression, have become areas of elevation, as in the instances of some coral islands, which, though formed during periods of depression, have been since raised above the waters to heights of from 200 to 300 feet or more.

---

[1] *Geology of India*, pp. 376 and 378.

[2] This has been contested. There are no doubt instances of reefs formed without subsidence, but for the depths of the Pacific Darwin's hypothesis best answers the geological conditions.

In conclusion, I may point to the imposing spectacle afforded by the slow secular upheaval of the vast tracts of Arctic lands on the shores of North America and Asia—an area of elevation so extensive that it embraces large portions of the land and islands bordering the Polar regions. This elevation has, in comparatively modern times, raised the land from 100 to 400 feet above the present sea-level, and is still, in our own times, in visible action over a superficial area extending in some directions, for thousands of miles.

[Some parts of the Fourth Article may be consulted in connection with this subject.]

VI

ON UNDERGROUND TEMPERATURES; WITH OBSERVATIONS ON CERTAIN CAUSES WHICH INFLUENCE THE CONDUCTIVITY OF ROCKS; ON THE THERMAL EFFECTS OF SATURATION AND IMBIBITION; AND ON A SOURCE OF HEAT IN MOUNTAIN RANGES, AS AFFECTING SOME UNDERGROUND TEMPERATURES.

§ 1. Introduction

As explained in the last paper the opinions of Physicists and Geologists as to what may be the probable thickness of the crust of the Earth differ very materially. On the strength of its great rigidity, Physicists contend for a maximum thickness of the crust, if not for the entire solidity of the globe. On the other hand Geologists contend, on the evidence of volcanic action, the crumpling and folding of the strata in mountain ranges, and its general flexibility down to the most recent geological times, for a crust of minimum thickness, and a yielding substratum, as alone compatible with these phenomena.

As we have considered the general question, my object now will be to give a fuller account of one branch of the subject—namely, the rate of increase of temperature beneath the surface,—a subject equally affecting the argument on both sides. My attention was specially directed to this subject in connexion with an inquiry commenced in 1850 on the temperature of Artesian wells, during which I found that the recorded observations gave so wide a choice in the selection of a mean rate for the increase of temperature, that very different values might be and were attached to that rate by different writers. I was thereby led to collect the very scattered evidence in England and abroad bearing on the subject, with a view to see whether it were not possible to fix upon some more definite and less elastic rate.

If the inquiry does not lead to any material alteration of the average now usually accepted, it may, at all events, serve to eliminate some sources of error, and also to restrict the limits within which the true rate may lie. It may also serve to show how far the differences in the observations in different geological areas are due to causes common to all, or how far they are special to each or due primarily to geological structure and local causes.

At present, owing to the wide differences in the recorded observations, it is very usual to take a mean rate of 1° F. for every 50 to 60[1] feet of depth, or of 1° C. for every 30 metres. Others take different rates of increase. These differences, I think, arise from taking too wide an average. Observations about which there is much doubt should be eliminated, and a selection made of a limited number of observations,

---

[1] A gradient of 1° in 50 feet has often been the one adopted by physicists and geologists, but one also of 60 feet to the degree, or even more, has been adopted on some authorities.

in which the sources of error have been best guarded against. These as I shall point out are very numerous. I had already in connexion with observations on Artesian wells, and as a member of the Royal Coal Commission of 1866, in connexion with the inquiry on the "Possible Depth of Working," got together a number of observations, and to those I have added a considerable number of others, many of them not before recorded. Although some of the earlier of these may seem to be of little value, I have thought it best to keep the record of all in the general list (Table I) for reference in case of need, as with corrected data respecting the surface temperature and height of ground some of them may possibly hereafter prove available.

I need scarcely say that I treat the subject solely from the geological point of view. For its mathematical and physical aspects, the papers of Sir William Thomson (now Lord Kelvin),[1] Professor Everett,[2] Professor Lebour,[3] and the Rev. O. Fisher[4] should be consulted.

---

[1] "On the Reduction of Observations of Underground Temperature," *Trans. Roy. Soc. Edin.* (1861), vol. xxii. p. 405; "On the Secular Cooling of the Earth," *ibid.* (1861), vol. xxiii. p. 157.

[2] "On a Method of Reducing Observations of Underground Temperature," *Trans. Roy. Soc. Edin.* (1861), vol. xxii, p. 429; "On Underground Temperature," *Proc. Belfast Nat. His. Soc for* 1873-4; Reports of Committee of Brit. Assoc. for 1868—1884.

[3] "On the Present State of our Knowledge of Underground Temperature," *Trans. North of England Inst. Min. and Mechan. Eng.* vol. **xxxi.** (1882), pp. 59, 204.

[4] *Physics of the Earth's Crust*, Chapter T.

The subject of the underground temperature attracted attention as far back as a century and a half ago, some observations having been made by Gensanne in 1740 in the mines of Alsace, which, if we take the mean annual temperature at the mines of 47° F., give, curiously enough, an increase of 50 feet per degree F.

Towards the end of the century, a few experiments were made by Saussure in Switzerland, and by Humboldt in America, but the more important ones were those made by Daubuisson in 1803, in the mines of Saxony and France, which gave a rate of increase varying from 54 to 72 feet per degree F. De Treba carried on similar observations in one of the Saxony mines for two years, which, taking the surface temperature at 44°, show an increase of 57 feet per 1° F.

Passing over some minor observations, we come to the series of careful and systematic observations commenced in the mines of Cornwall by Dr. John Forbes and by Mr. R. Were Fox about 1820, and carried on continuously until 1857. Other Cornish geologists followed. Mr. W. J. Henwood in particular made, between 1837 and 1858, a large number of experiments, not only in Cornwall, but also in the mines of South America. Notwithstanding that the instruments had not the perfection of those of later date, the observations are of much value, as they were conducted under very varied conditions, and with a full understanding of the various causes of interference to be guarded against. They show very considerable variations in the rate of increase with depth. Mr.

Fox records a range of rate for 1° F. of from 32 to 70 feet, from which De la Beche deduced a mean from 46 to 51 feet; and Henwood, while finding the results to range as widely as did Mr. Fox, nevertheless estimated from a large number of averages the gradient for 1° F. to be from 37 to 41 feet only.

In 1822, Cordier published his celebrated "Essai sur la Température de l'Intérieur de la Terre," and in this he recorded the observations previously made by Gensanne, Daubuisson, and Fox, together with some observations made by himself in the coal-pits of France. The several results showed variations of rate extending from 13 to 57 metres per 1° C., but he concluded the mean to be about 1° C. per 25 metres of depth.

Amongst the most careful of the early observations are those made by De la Rive and Marcet in an artesian well at Pregny near Geneva. In this case, the thermometer was protected against pressure, and the result gave a rate of increase of $48\frac{1}{2}$ feet per 1° F.

This was shortly followed by Professor Phillip's observations in the deep coal-pit of Monkwearmouth, Sunderland; great precautions were taken against error, and in the result an increase of 1° for every 62 feet of depth was obtained.

At this time a number of deep artesian wells were being made in France by Messrs. Degousée and Laurent and other engineers, and M. Walferdin invented an overflow maximum thermometer, which, carefully guarded against pressure, gave excellent results, and which he employed to check many of the earlier observations made without this necessary precaution. Others also were made about this time in Algeria, Venice, and other parts of the Continent by Degousée and Laurent. Those in Algeria seem to show that the rate of increase is there more rapid than in Europe.

Admirable discussions of the question were made from time to time by Arago, and finally in his *Œuvres Complets* in 1856. In this latter he reviews the whole subject, gives elaborate particulars of most of the re-recorded observations, with the advantage of more accurate surface temperatures, and concludes that the rate of increase is very variable, but may be taken as a mean at from 20 to 30 metres for each degree Centigrade.

In 1861 Sir W. Fairbairn gave an account of the observations, so frequently referred to, in the Dukinfield Colliery, where the rate of increase was estimated at 1° F. for every 79 feet.

Professor Hull also brought together in his *Coal Fields of Great Britain* (Edit. 1873) a number of observations made in the coal-pits of this country, and drew especial notice to the Rosebridge Colliery temperatures. The general conclusion at which he arrived was that the underground temperature varies, being much influenced by the dip of the strata, but may, as a mean, be taken at 60 feet per 1°.

The Royal Coal Commission of 1866-70 was the means of obtaining through its Committee "On the Possible Depth of Working," a mass of valuable evidence from practical coal-owners, inspectors, and managers, relating to the temperature at depths in the coal mines of Great Britain. The general opinion of the witnesses was that the temperature increased at the rate of about 1° F., per 60 to 63 feet of depth, and the Committee concluded that the rate of increase of the temperature of the strata in the Coal districts of England was in general about 1° F. for every 60 feet of depth.

In 1867 a Committee of the British Association was appointed "for the purpose of investigating the Increase of Underground Temperature downwards in various Localities of Dry Land and Under Water," with Professor Everett as secretary. To this Committee we are indebted for a series of sixteen valuable annual reports, in which are recorded a large number of observations carried on with corrected instruments supplied by the Committee, and under the able superintendence of its secretary. Each set of observations is discussed at length; the causes of interference are considered; and detailed particulars are given of all the conditions of depth, strata, isolation, &c., under which they were made.

In the summary of results given in the Report for 1882, Professor Everett, after describing the instruments used, the methods of observation, and the many questions affecting the value of the observations, such as the heat generated by the boring tools, or by chemical action, the effects of ventilation, convection currents, &c., proceeds to a comparison of results. Taking the list of 36 localities of which Professor Everett gives the recorded results, and classifying them as—1. Metallic Mines; 2. Coal Mines; 3. Wells and Wet Borings; 4. Tunnels; he concluded that in the—

Metallic Mines, the increase of temperature with depth varied from 1° F. in 47 feet to 1° in 126 feet;

Coal Mines, the range was 1° F. in 45 feet to 1° in 79 feet;

Wells and borings gave from 1° F. in 41 feet to 1° in 130 feet, or excluding the wells of La Chapelle and Bootle, which are open to suspicion of convection currents, the least rate of increase was 1° in 69 feet.

Tunnels. In the only two great tunnels in which observations have been made, that of Mont Cenis gave an increase of 1° for 79 feet, and that of St. Gothard of 1° for 84 feet, but in a subsequent report Dr. Stapff has shown that an important modification is necessary for a portion of the tunnel.

In deducing a mean from these various results, Professor Everett considered it best to operate not upon the number of feet per degree, or what he has termed the thermometric gradient, but upon its reciprocal—the increase of temperature per foot; and assigning to the results of the observations at the thirty-six localities weights proportional to the depths, he found the mean increase of temperature per foot to be 0·1563, or about $\frac{1}{64}$ of a degree per foot—that is, 1° F. in 64 feet.[1] In the subsequent report of 1883 however, taking the corrected readings of Stapff for the St. Gothard and Mount Cenis tunnels, this mean is reduced to 1° F. in 60 feet.

Observations have been taken on the deep hot artesian well of Buda-Pesth, by Professor Szabo. Others have been made on some deep wells in America by American geologists; while temperatures of a number of ordinary wells have been given by the Rivers Pollution Commission. Amongst those to whom I

[1] Professor Everett observes, however, in a previous paper (1874) that some of the best observations give a rate of about 1° F. for 56 feet of descent, and that this seems a fair average.

am indebted for furnishing me with observations in wells and mines in their respective districts, not hitherto recorded, are Professor G. Dewalque of Liège, Professor Gosselet of Lille, M. F. Cornet of Mons, and others.

Professor Lebour has further discussed the subject, with reference especially to the nature of the experiments still required to improve our knowledge of it, in an interesting paper read before the North of England Institute of Mining and Mechanical Engineers, in 1882. In this he gives a list of observations made at fifty-seven localities, and deduces from them an average rate of increase of 1° F. per 64·28 feet of descent; but he remarks that all the observations recorded are by no means of equal value, and points out that the mean rates of increase in the best observations cluster about the numbers 1° F. per 50 or 60 feet.

§ 2. General List of Observations in Order of Date. (*Table I.*)

I have given in a general list all the observations on underground temperatures of which I have been able to find a record.[1] The number of different localities where such observations have been recorded amounts to 231, and the number of stations[2] to 530.

[1] I should feel greatly obliged by information respecting any other observations.

[2] By "station" I mean all the separate points in a mine, or at the several depths in a well, at which the observations were made.

I have only omitted a few which were obviously wrong, together with some of the observations repeated frequently in the same mine, or repeated more frequently than necessary for our object in mines in the same district, and which would only add needlessly to the length of the list.[1] Neither have I considered it necessary to give in every case all the details, as these will be found in the original papers referred to, but have confined myself to giving the principal and essential facts and results.

The observations are given in the order of date, an order which, as a general rule, agrees with that of their reliability, although there are several remarkable exceptions. The superiority of the later observations consists chiefly in the perfection of the instruments and methods of experimentation, although most of the more general disturbing causes interfering with accuracy of the results were understood and guarded against, more or less well, by the early observers.

The particulars of depth, temperature, and modes of proceeding, are given in the terms of the original observers. On two points only, on which their information was often unavoidably imperfect, have I made any alteration—points which even now are in many cases not accurately determined—viz., the mean annual temperature of the surface, and the height of the surface above the sea-level. These alterations will account for occasional differences in

[1] Mr. Henwood's long and valuable series of observations in the Cornish and foreign mines will be found in vols. v. and viii. *Trans. Roy. Geol. Soc., Cornwall.* I have taken only the more certain of these, omitting duplicates.

the *thermometric gradients*, between the observations as originally recorded and as given in these lists, and will in most cases be found referable to my taking a different mean annual temperature to that used by the original observer.

Height of the Ground.—As this influences both the surface and underground temperatures, it is important that it should be determined with precision. This is not always possible, although the later publications of the Ordnance Survey have in many cases furnished us with levels, which, if not at the precise spot, are sufficiently near to form, with a knowledge of the country, a near estimate; but not unfrequently mines are in out-of-the-way places, where only roughly approximate estimates can be made. Several of the surface heights given in this table necessarily come into this category, and are therefore subject to future correction; while there are many cases in which I am unable to give even an approximate height.

Mean Annual Temperature.—This is a point on which it is very desirable to have exact information, as an error even of one or two degrees materially affects the calculation. There are yet, however, many of the places in the lists, where satisfactory information is wanting, and the inserted temperature may require revision. For most places on the Continent, the works of Arago,[1] which embody all that was then known of their mean temperature, both in France and other parts of the world, are followed. The publications of the Meteorological Society of

[1] *Notices Scientifiques,* vol. v, p. 518. Edit. 1857.

Scotland have done much to give us exact data respecting the mean temperature of places in Great Britain,[1] and I am further indebted to Mr. Robert H. Scott, F.R.S., of the London Meteorological Office, for the latest recorded mean temperature of many places both in this country and on the Continent. There are, besides these, the older and more general tables of Dove.

Where the temperature of the locality cannot be had, that of some place near, and on or near the same level, has been taken. Where there is an important difference of height between the place at which the surface temperature has been taken and the site of the underground experiment, an approximate allowance of one degree F. (+ or −) is made for every 300 feet of difference of level. When, however, as is sometimes the case, a mean annual surface temperature is not obtainable, a near approximation may often be made by taking the temperature of ordinary surface wells when about 40 to 50 feet deep, and of springs. This plan has been, especially in the earlier observations, frequently adopted.

What is really desirable is that the datum line whence to commence the measure of the increase of temperature with depth, should be placed at that point below the surface, at which the changes of annual temperature cease to have effect. Any such determination[2] might also with advantage be applied to the correction of the older observations.

[1] Especially the Journal for November, 1883.

[2] This was done in the case of the experiments at Grenelle, where a datum line based on the temperature and depth of the

In the absence of this information the gradient has sometimes been calculated from the surface, and at other times it has been calculated by supposing the mean invariable surface-temperature to lie at a depth of 50 feet, and taking that as the datum level from which to start.

Either, therefore, some of these estimates may be too high in consequence of the deduction of 50 feet from the depth, or others are too low in consequence of not making an allowance for the zone of variable temperature. But as the mean annual temperature of the ground at the surface generally in these latitudes exceeds that of the air by about 1°, it follows that if we take the mean temperature of the air at the place of observation, and calculate the rate of increase from the surface, instead of allowing 1° for the higher temperature of the ground, and placing the datum line 50 feet lower, we shall come to nearly the same result. At the same time it must be admitted that, in so doing, we may sometimes be liable, especially with stations of moderate depths, to make the rate of increase with depth less than it should be. It is therefore the more important in future observations to determine if possible the temperature at a depth of 50 to 60 feet, or where the first bar of uniform temperature may happen to be, and to calculate from that point the rate of increase of temperature with depth.

cellars of the Paris Observatory was employed. These are 29 mètres (95 feet) deep, and the thermometer stands invariably at 11·7° C. (53° F.), the mean annual temperature of Paris being 10·6° C. (51° F). This is, however, a greater depth than needed.

In looking over Table I, the marked differences respectively in the results obtained in Mines, Coal-pits, and Artesian wells will be apparent. This arises both from the circumstance that not only in each of these are the geological conditions very dissimilar, but also that the disturbing causes are of a different order. In Metallic Mines, the latter are attributable to—

1st. Air currents established for ventilation;
2nd. The circulation of underground waters;
3rd. Chemical reactions;
4th. The working operations.

And in Artesian Wells, to—

1st. The pressure of the column of water on the thermometer;
2nd. Convection currents in the column of water.

While a general cause affecting each group is *conductivity*, which differs in the various rocks, and which is itself liable to be influenced by a number of causes to which we shall refer separately.

We shall therefore divide Table I into groups in accordance with these conditions, and take each separately in the following order:—

1. Coal Mines.
2. Mines other than coal.
3. Artesian Wells and Bore-holes.
4. Tunnels.

§ 3. OBSERVATIONS IN COAL MINES. (*Table II.*)

Considering the general uniformity in geological structure of the Coal-measures, the temperature observations in Coal Mines are more discordant than might be expected. This depends upon various causes, of which Ventilation is seemingly the principal, while in some cases differences of Conductivity may have influenced the results. It is true that in most cases the sources of error have been guarded against, but it is doubtful whether sufficient allowance has always been made for them. The main disturbing causes are as follows:—

Loss of Heat through Exposed Surfaces.—To guard against the cooling of the Coal or Rock produced by ventilation, it has been customary to place the thermometer in holes 2 to 3 feet deep, but it is a question whether this is sufficient, for we are rarely informed, as is essential, of the precise time that the face of the coal or rock has been exposed. In most observations we are only told that they have been made in "freshly exposed faces." But as on these working faces the ventilation is necessarily well maintained, the difference between the temperature of the air and of the strata is there often very considerable, and therefore the cooling of the rock very rapid.

On the surface of the ground the diurnal variation of temperature extends to the depth of about 3 feet; and it has been further shown by means of thermometers sunk in the ground, that in the neighbourhood of Edinburgh at a depth of [1]—

1 foot the monthly fluctuation is from	33° to 54·0°	or =	21·0°
2 feet	,,	,,	,,	36 to 52·5	or = 16·5
4 ,,	,,	,,	,,	39 to 51·8	or = 12·8
8 ,,	,,	,,	,,	42 to 50·0	or = 8·0

[1] Prof. James D. Forbes, *Trans. Roy. Soc. Edin.*, vol. xvi. p. 189.

At the depth of 29 feet at Greenwich, the annual fluctuation is still 2° to 3° F., and to reach the limits of range, or the plane of uniform annual temperature, we must in these latitudes go to a depth of from 50 to 60 feet.

Owing to ventilation, there is a permanent difference between the temperature of the air in a mine and the temperature of the rock or coal, analogous to the diurnal variation on the surface of the ground, and any such surface of rock exposed to air of a lower temperature will lose heat with a rapidity proportionate to the difference in temperature between the two bodies. Therefore as convection currents and ventilation are constantly introducing into the shaft and mine air of a lower temperature than the underground rocks, these latter must suffer a loss of heat from the moment they are so exposed. To avoid this, the best observers have operated on freshly exposed surfaces, where the loss is at a minimum; on the other hand, numerous observations have been made on surfaces exposed for a length of time and often under conditions of which we are not informed.

In the case of the Dukinfield Colliery, near Staleybridge, it is stated that the bore-holes were driven to such a depth into the side of the shaft as to be unaffected by the temperature of the air in the shaft, and the thermometers were left in the holes from $\frac{1}{2}$ to 2 hours; but we are not informed of the depth of the holes, nor of the temperature of the air in the shaft, nor of how long the rock had been exposed. The shaft, which was a very large one and carried to a depth of 2151 feet, took 10 years to sink. At this depth the recorded temperature of the rock was 75° F., and the estimated rate of increase was calculated to be 99 feet for 1° F. The observations were carried on during the whole of the time that the work proceeded. Although we do not know the temperature of the air, the action of convection currents or of ventilation is clearly shown in the irregularities of the rock temperatures in summer and winter, even at great depths. Thus, on the 12th June, 1849, the shaft had reached a depth of 704 feet, and the rock temperature that depth was 58°. The same temperature was noted with slight fluctuations until the 22nd December, at which time the depth had reached 810 feet. In February, 1857, at a depth of 1450 feet, the temperature was $67\frac{1}{4}°$, which rose to $72\frac{3}{4}°$ in August, at a depth of 1689 feet. It then slightly decreased, partly recovering and ending in October at $72\frac{1}{4}°$, when the depth was 1840 feet. In March, 1858, it had fallen to $71\frac{1}{2}°$ at 1881 feet, and then rose by July, at 2055 feet, to $75\frac{1}{2}°$. After the winter, in March, 1859, at 2151 feet the thermometer stood at 75°, or half a degree less than it did in July at a depth of nearly 100 feet less. I can only attribute these fluctuations to the influence of the seasons and convection currents, or to some unmentioned ventilation. It is evident that such records can furnish no true test of the real "thermometric gradient." A new shaft sunk in 1858, not far from the other, gave, in the same way, analogous results, viz., an increase of 1 F. for every 90 feet.

When, however, at a subsequent period, and when the works of this colliery had been carried to the great depth of 2700 feet, observations were made in galleries at a distance from the shaft, with instruments furnished by the British Association Committee, and in holes 4 feet deep, an amended reading of 1° for every 72 feet was obtained. Still that rate of increase is slow. Professor Everett has, however, drawn attention to the fact that there are other collieries, such as those of Ashton Moss, Denton, and Bredbury, within a few miles of Dukinfield, in which the results are in close agreement with those obtained in the latter. It has been suggested that the rapid dip of the strata at Dukinfield facilitates the transmission and escape of heat. It is quite possible that this cause may have an influence, as the experiments of Herschel and Lebour have shown experimentally that the conduction of heat is more rapid along the planes of cleavage, or bedding, than across them. Nevertheless no such difference exists in other mines where the stratigraphical conditions are very similar; the strata in the Liège, Mons, and Valenciennes coal-fields (Nos. 140, 214-16, 145) are more disturbed, and dip at more rapid angles than at Dukinfield, and yet the deep temperatures there show no exceptional gradients.

Mr. Garside also draws attention to the circumstance that the Red Sandstone, there overlying the Coal-Measures is highly charged with water, which, as it is largely drawn upon for water supplies, may possibly affect the temperature conditions.

Some of the anomalous results may be attributable

to the excellent ventilation of these deep pits. At the Dukinfield Colliery, about 58,000 cubic feet of air circulate through the pit per minute. The mean annual temperature of the air at the surface is 48° F., while the return air from the pit has a temperature of 73°. But then again at Rosebride and other large pits—where the rate of the increase of temperature is more rapid and nearer the probable normal—the ventilation is not less carefully attended to. Still, that ventilation has a preponderating influence is shown by an experiment of Mr. Wynne, who found that in a rise of the Dukinfield Colliery, where owing to defective ventilation the temperature of the air and rock were both 76°, the rate of increase from the surface gave 55 feet per degree.

Other causes have also been suggested for the lower temperature of the Dukinfield pit. Mr. Inspector Dickinson states that before the shaft was sunk, two of the principal upper seams of coal had been worked away from other pits, and that the lower seams had been worked a long way from the outcrop down towards the Astley pit.[1]

The Dukinfield observations, therefore, while they confirm the general law of increase of temperature with the depth, are scarcely admissible in estimates concerning the exact rate of increase, and it will be safer to omit them and all similar observations in calculating the mean.

Though good observers will avoid long exposed and cooled surfaces, it is evident that as cooling commences at once, it is difficult to escape, however early

[1] *Coal Commission Report*, vol. ii. "Depth of Working."

the experiments are made, from the effects of ventilation. It is not a question of weeks or months, but of days.

The Effects of Ventilation.—The evidence given before the Committee of the Royal Coal Commission of 1867, "On the Possible Depths of Working," is particularly valuable for the information it affords respecting the effects of ventilation.

At the Pendleton Colliery (Nos. 104 and 122), near Manchester, the coal in a level 500 yards from the down brow had a temperature of 70°, while at 1000 yards distant, in the same level, when freshly opened, it had a temperature of 83° F.

At the Rosebridge Colliery, Wigan, the temperature of the Raven coal in the shaft was 78°, whilst at 630 yards distant in the pit it had a temperature of 88°. That of the air at the bottom of the down-shaft at a depth of 1800 feet was $60\cdot5°$, and, after circulating 1500 yards, 73°. At another time it was $51\cdot5°$; after circulating 2400 yards, $67\cdot5°$; and after 3140 yards, 71°. In July, 1858, the strata at the same depth had a temperature of 80°; in March, 1861, it was reduced to 72°.

At the Annesley Colliery (No. 106), Nottinghamshire, a fresh cut coal at a depth of 1425 feet had a temperature of 73°, and six months later it had fallen to 64°. In another case, the coal cooled 11° in three months.

The volume and velocity of the air currents vary greatly. In one place a circulation of 5000 cubic feet per minute is recorded. In another pit the circulation was 100,000 cubic feet per minute, while in the upcast shaft, 7½ feet in diameter, of another pit, 150,000 cubic feet of air passed per minute.

At the Hetton Colliery (No. 116), Durham, the dependence of the temperature of the strata upon distance from the shaft, in connexion with ventilation, is well shown in a table by Mr. Wood, of which the following is an abstract.

Surface temperature at date = 36°. Holes in coal 3 feet deep.

Depth from surface.	Distance from shaft.	Cubic feet of air per minute.	Temp. of coal.	Temp. in current of air.	Rate of increase in depth for 1° F.
feet.	yards.				feet.
1,100	312	104,000	60°	50°	100
1,270	915	50,000	63	58½	91
1,360	1,640	18,000	66	62	80
1,354	3,664	3,200	68	68½	67
1,400	3,550	4,972	70½	72	64
1,395	4,332	5,000	71	73	63½

In this case we have four observations made nearly at the same depth, and yet showing a rate of increase with depth varying from 80 feet to 63½ feet for 1° Fahr.

In the Seaham (Durham) Colliery the circulation of air is very large,—193,346 cubic feet per minute. At a depth of 1524 feet and 1642 yards distant from shaft, the temperature of the air was found to be 61°; at 2726 yards distant and at the same depth it was 67°; and at the face of the coal 82°. The temperature of the return air = 76½°.

At the Eppleton Colliery, Durham, in October (?) the temperature of the air at the intake at a depth of 850 feet and 110 yards distant from shaft was 44°, and after circulating 1960 yards, $59\frac{1}{2}°$, or an increase of $15\frac{1}{2}°$.

Some joint observations by Mr. Wood and Mr. Dickinson in the Monkwearmouth Colliery (No. 30), Sunderland, gave the following results. 29th March, 1870. Temperature of surface = 42°.

Depth.	Distance from shaft.	Temp. of the air.
1676 feet	Bottom of down shaft	50° F.
1676 ,,	2134 yards	66 ,,
1638 ,,	2850 ,,	74 ,,
1646 ,,	3112 ,,	81 ,,
1640 ,,	4030 ,,	82 ,,

This pit is so dry that it requires water to lay the dust.

Mr. Dickinson found that at a depth of 2088 feet in one of the Pendleton pits the temperature of the coal at 500 yards from the engine brow was 70°, and at 1000 yards distant 82°. In the same pit at a depth of 2214 and distant 200 yards from the engine brow, the coal in a hole 3 feet deep was 76°, and at a distance of 930 yards 82°, while in a hole $7\frac{1}{2}$ feet deep the temperature was 84°. The general opinion amongst the witnesses seemed to be that the effect of ventilation was to reduce the temperature of the pits some 9° to 10° F., though some estimated it at as much as 15° to 20°.

Mr. Lupton mentions that in a pit 1425 feet deep the fresh cut coal had a temperature of 73°, whereas at the end of six months it had fallen to 64°. In another case a period of three months had sufficed to reduce the temperature 11°.

After the air in the pits has circulated a certain distance (estimated by one of the witnesses at one mile), it sometimes becomes in the face of the workings hotter than the rock; the difference (2°) being due to the heat of the lights and the men.

The extent to which the temperature of the air affects that of the coal is well shown in the table (see next page) relating to the Durham Collieries put in by by Mr. L. Wood. In one case (Report, p. 140), the

Jane Pitt, Hetton Colliery. Depth below surface.	Circulation of air per minute.	Distance from shaft.	Temperature of				Temperature— taking the mean surface temperature at 47°.	
			Intake air.	Coal.	Return air.	Coal.	1° for 60 feet of depth.	1° for 50 feet of depth.
feet.	cub. ft.	yds.						
1,080	41,800	410	58½°	65°	70½°	68°	65°	68½°
1,270	41,580	955	59½	65	71	73½	68	72½
1,320	30,030	1,247	62	66½	69½	72	69	73½
1,360	28,000	1,640	63	67	70	72½	70	74¼
Average 1,357½	35,352	1,063	60¾	65¾	70¼	71½	67½	72¼

difference between the coal and the air in the part of the pit through which the intake air circulates, was from 4° to 6½°, or an average of 5°; whereas in the return air at the same distances from the shaft, the differences were reduced to an average of 1¼°.

In the Crumbouke mine the face of the coal which had been worked two years had, according to Mr. Dickinson, lowered 4°. A similar loss was shown at the Rosebridge Colliery, but after a longer interval. In the adjoining pit, the difference between the temperature of the air in the galleries and of the coal varied from 10° to 12½° F.

At the Rosebridge Colliery other observations made three, four, and nine years after opening the colliery showed, that at 1410 feet deep, and 200 yards from shaft, the temperature of the coal was = 66° F., and at 360 yards = 70° F., while at 1800 feet deep and 215 yards from shaft the temperature of the coal was = 65° F. Mr. Bryham states that the reduction of temperature effected in this pit by ventilation amounts to from 15° to 24° F., or even more in a strong current.

Other returns give the column of air circulating through the pits at from 50 to 150 million cubic feet in the twenty-four hours, with a gain of temperature of from 10° to 20° F.

The following is an abstract of Mr. Lindsay Wood's tables :—

	Depth below surface.	Distance from downcast shaft.	Intake air, cubic feet per minute.	Temperature of	
				Intake air.	Return air.
	feet.	yards.		Fahr.	
Eppleton Colliery, Durham, 13th Oct., 1869. Mean surface temp. = 53° F. Working face...	1,100 1,450 1,395 ,,	312 2,925 4,130 4,440	78,720 11,550 4,500 ,,	61·0° 69·0 72·0 74·0	70°
The same pit *off work*, 23rd Oct., 1869. Surface temp. = 52° F.	1,100 1,395	312 4,130	52,200 3,240	57·5 71·5	69
Monkwearmouth Colliery, 25th Oct., 1869. Surface temp. = 45° F. Working face...	1,676 1,646 1,638 1,646	80 1,430 2,850 3,256	38,250 71,500 5,000 ,,	56·5 63·0 75·7 81·2	77
Murton Colliery—*off work*. Working face... Surface temp. = 50°.	1,339 1,374	410 4,400	76,160 1,600	54·0 70·0	68

Even in the two pits off work, we have in the one instance 44,550 cubic feet of air passing out per minute, with a gain in temperature above that of the outer air of 17° F., and in the other case 25,500 cubic feet pass out with a gain of 18° F.

In these instances the temperature of the outer air was only in one observation under 50° F. At other seasons of the year, with the outer thermometer lower, and with the difference of temperature between the circulating air and the strata greater, the loss of heat would proceed still more rapidly. Mr. Forster mentions that he has known ice form at the bottom of a coal shaft, and Mr. Bald, speaking of the Whitehaven Collieries, states that the circulation of air sometimes causes the water to freeze on the sides of the shaft, and "even form icicles upon the roof of the coal within the mine," whilst the air from the rise-pit issues in a dense misty cloud.

In consequence of the allowance necessary to be made for these effects of ventilation, some of the witnesses, while taking the actual rate of increase at about from 50 to 60 feet for each degree F., considered that the normal rate might not be more than 50 feet per degree.

As a rule, the deeper the mine the greater the ventilation, and therefore, the more rapid must be the cooling of the underground strata. But the amount of ventilation depends also not only on depth, but on the quantity of gas present in the coal. The late Sir Warington Smyth said,[1] that "in round numbers 100 cubic feet of air per minute may be required for the health and comfort of each person underground, or for 100 men, 10,000 cubic feet; but if fire-damp

[1] *Coal and Coal Mining*, p. 205.

be given off—say at the rate of 200 cubic feet per minute—we should need at the very least thirty times that amount of fresh air to dilute it, or 6,000 cubic feet in addition. Increase the number of men and liability to gas, and 40,000 to 60,000 cubic feet of air may be indispensable for safety."

It appears, therefore, that the rate of cooling from ventilation in different coal mines and at different depths is very variable, and that this may in part account for the great discrepancies between the thermometric gradients at different collieries. Unless, however, we are in possession of all the collateral conditions, it will not be possible to assign in each instance the precise weight to be attached to this disturbing cause.[1]

Other Causes which may affect the Temperature of the Coal Strata in Underground Workings.—Apart from the chemical decomposition of minerals, which is of rare occurrence in coal mines, and the heat arising from the men, horses, and lights, which can only be of importance in shallow mines, for in deep mines it is rarely that this source of heat brings up the temperature of the air to the normal temperature of the rock, there are few causes likely to produce an abnormal rise of temperature, except one which, though of exceptional occurrence, should be noticed in connexion with this subject. This is the heat produced by the crushing of the rocks which sometimes takes place in coal mines.

[1] Sir W. Warington Smyth gave, in his Anniversary Address to the Geological Society in 1868 (pp. 79—87), some interesting particulars respecting ventilation and other causes affecting underground temperatures.

The Crushing of Rock.—It is certain that in coal mines when the pressure crushes the pillars and forces the strata up in *creeps*, more or less heat is liberated. No experiments have been made to determinate what, in these cases, are the exact effects produced, though some of the evidence given before the Coal Commission sufficiently establishes the general fact. Mr. Elliot,[1] in speaking of one of the pits in South Wales mentioned that when "a creep takes place, he has known the temperature very much increased," and in one case where "the pressure began to crush the pillars, the heat produced was so great that he feared it would set fire to the coal." In some cases the pressure has been such as "to grind the rock to powder, like the effect of a dozen mill-hoppers, and this was accompanied by considerable heat." He had often found the air very hot when a sort of temporary creep occurred.

Escape of Gas.—On the other hand, a cause productive of a loss of heat is a more constant disturbing cause. There are few coals which do not give off gas when first exposed. In some seams it may be observed exuding from the freshly broken surfaces with a hissing sound; and, if in large quantity, as in the case of the so-called "blowers," or sometimes near faults, it issues with a rushing noise like the steam from a high pressure boiler. Some of these blowers will be exhausted in a few minutes, others will last for years, like the one at Wallsend, which gave off 120 feet of gas per minute.[2] The common gas on these occasions

[1] *Coal Commission Report* "On Possible Depths of Working," p. 112. [2] *Coal and Coal Mining*, p. 204.

is light carburetted hydrogen, which, under the enormous pressure, must exist in the coal in a highly condensed, if not in a liquid state, otherwise it is hardly conceivable how the discharge could be maintained so long. The elastic pressure of this gas itself is said sometimes to equal 300 to 400 lbs. to the square inch.[1]

In any case the escape of this gas from the coal, in which it seems to be stored in an infinite number of minute cavities, must be to reduce the temperature of the coal from which it escapes, though here again no special experiments have been made to ascertain the exact loss of heat from this cause. I find, however, one of the witnesses [2] on the Coal Commission remarking that in a pit the depth of which had been increased from 830 to 1588 feet, the temperature of the coal was lower at the greater than at the lesser depth, and he attributed this to a strong blower of gas issuing from the coal, which at that point was sensibly cooler to the touch.

On a table also put in by Mr. Elliot, and referring apparently to the same case, it is stated that—

In the Lower Duffryn Mine at a depth of...... 1588 feet,
—and distant from shaft...... 1850 yards,

the temperature of coal in a wet hole with a blower of gas was 62° F., whereas it was found that—

At a depth of...... 1269 feet,
Distance from shaft............ 1020 yards,

---

[1] Sir Frederick Abel states that, if cavities are bored into the coal and plugged, the gas speedily accumulates to such an extent as to exercise a pressure of several hundred pounds upon the square inch, as indicated by pressure-gauges fixed into the cavities. *Nature*, December 3rd, 1885.

[2] Mr. Wilmer: *Coal Commission*, p. 118.

the temperature in a dry hole with little gas was 74° F.

At the Cym Neol Colliery, at depths of 990 and 1150 feet, and 1350 and 1070 yards distant from shaft, the temperature was respectively 62° and 65°, which is abnormally low, but it is stated that the observations were made in wet holes *with blowers*.

This escape of gas from the body of the coal may account for the observation of another witness to the effect that " the coal gives out heat quicker than the rock," and that the temperature of the coal is, on the whole, less than that of the associated rocks. In several cases observations made in the same pit, both in the coal and in the underlying shale, and generally the temperature of the rock has been found to be higher than that of the coal. In the Rams Mine, Pendleton, for example, the temperature of the coal in a hole 3 feet deep was 82° F., and in the floor 83° F.; in the Crumbouke Mine the coal was 80° F., and the floor 82° F. In other levels it was 80°, and 78° against 82° F. In one of the Ruabon collieries, however, the relative temperatures varied, from some unexplained cause, as follows :—

Depth 1002 feet; temperature = 60° F. in coal, and 67° F. in floor.
,, 1503 ,, ,, 70½ ,, 68 ,,
,, 1605 ,, ,, 73 ,, 77 ,,
,, 1770 ,, ,, 78 ,, 74 ,,

As the conductivity of the coal (0·00068) is less than that of sandstone (0·00672) or shale (0·00235), the coal should retain its heat longer than the rocks. If the coal is cooler than the rock, it must arise from an independent source of cooling in the coal, such as this of the escape of gas condensed under great pressure. It is an effect important to notice, inasmuch as not only is the discharge of gas from coal of almost constant occurrence, but also from the circumstance that it affects especially the fresh opened surfaces of coal in which the temperature observations have so generally been made.

The Effects of Irregularities of the Surface.—Although it was known that the underground isothermals under mountain masses are not prolonged in horizontal planes, but follow in a certain ratio the curves of the ground above, little was known to what extent mines were affected by the smaller irregularities of the surface before Mr. George Elliot's observations in the coal pits of South Wales. These show that there is a sensible increase of heat with an increase in the mass of the superincumbent strata—a point that has rarely been sufficiently taken into consideration in underground temperature observations.

His observations[1] were mostly made in abandoned parts of the pits, so as to avoid as far as possible both the effects of ventilation and of working, and the holes, in which the thermometers were placed, were 4 feet deep. Only the depths and temperature of the coal are given in the original sections. I am responsible for the isothermal lines which I have plotted on those data, for the purpose of showing the more precise manner in which the temperature at depths is influenced by the surface levels.

[1] *Coal Commission Reports*, vol. ii., p. 105—111.

At Upper Duffryn Colliery the top of the shaft is 400 feet above the sea level, and the shaft 360 feet deep. The seam of coal has been followed southward for a distance of 2,327 yards, at which point (No. 4)

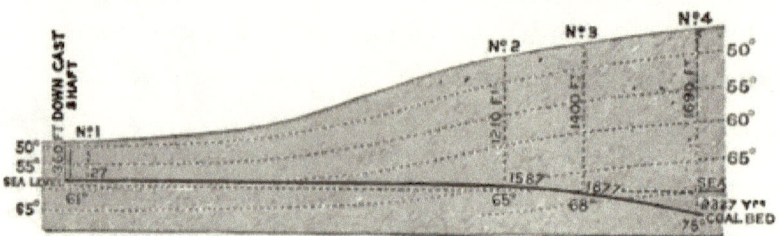

Fig. 1.—Section of Upper Duffryn Colliery, Aberdare District.

it is 190 feet lower in respect to the sea level than at the entrance. The distance from the shaft might account for an increase of 3° to 4° in the temperature, but not for the difference of 14°. This is due to the circumstance that at this point the surface of the ground has risen so much that the seam of coal lies 1,690 feet below the surface, or 1,330 feet deeper than at the bottom of the shaft.

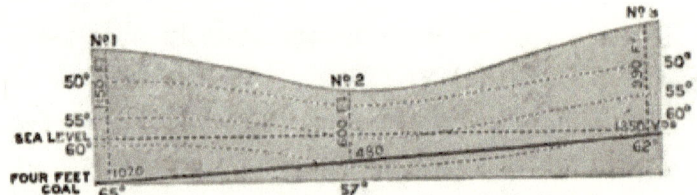

Fig. 2.—Section of Cym Neol Colliery, Aberdare District.

At the Cym Neol Colliery one station (No. 2) is under a valley, and two (1, 3) under the adjacent hills, rising about 500 feet above the valley. No 2 station is only 480 yards distant from shaft (not in section), against 1,070 from Station No. 1, and

1,350 yards from Station No. 3. The curves of temperature are very apparent. Outside air 69° F.

In a similar way the New Tredegar Colliery is situated in a valley 670 feet above the sea level, and the works pass under a hill which rises 617 feet

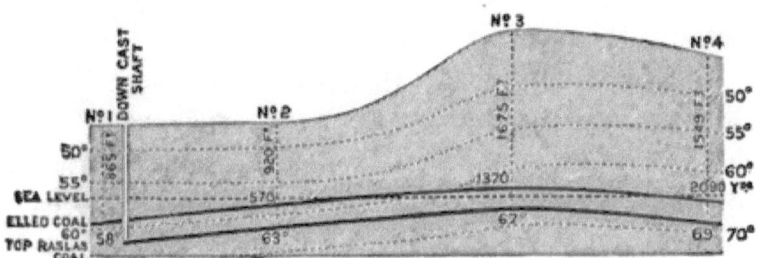

Fig. 3.—Section of New Tredegar Colliery, Aberdare District.

above the valley, with the effect of raising the temperature of the coal 11° F., though a portion of this may be owing to distance from shaft. The thermometer outside stood at 60° to 70° F.

At the Vochriw Dowlais Colliery, on the other

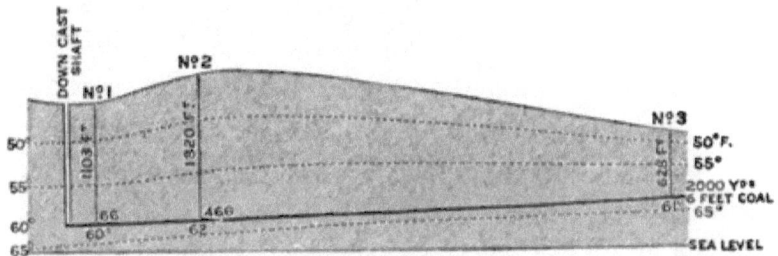

Fig. 4.—Section of Vochriw Dowlais Colliery, Aberdare District.

hand, the top of the shaft is on a hill 1,330 feet above the sea level, and the seam of coal, which there lies 1,103 feet below the surface, is followed first under a hill which rises 217 feet above the pit mouth and then under a valley only 818 feet above sea level, or 512 feet lower than the pit mouth. Notwithstanding the distance of Station No. 3 from the shaft, the fall of temperature due to the lowering of the surface is apparent, though the temperature at Station No. 2 is too low, which the witness attributed to a strong current of air passing at the time. Outside the thermometer marked 59° F.

It will be noticed that the observations in all these pits were made when the thermometer outside stood high, so that the cooling effects of the outer air would be less at the stations near the shaft than at other seasons.

Amongst other pits where recorded temperature observations have been made, and which are situated in districts sufficiently hilly to be affected by irregularity of surface, are the Dukinfield and some of the other pits in the Manchester district.

Selection of Observations.—It will be apparent from the existence of these disturbing causes, that not only ought we to know that the instruments employed are properly constructed and placed, but we should also know,—1st. Height of the ground above the sea level. 2nd. The mean annual temperature of the surface. 3rd. Depth beneath the surface of the ground of the stations. 4th. Distance of those stations from the shaft. 5th. Temperature of the air and volume in circulation. 6th. Length of time that the face of the coal or rock has been exposed. 7th. Whether or not there is much gas given off from the coal. 8th. Depth of trial hole and whether in the coal or rock. Besides these more essential points, the temperature of the air outside, the dip of the strata, their hydro-geological conditions, together with any local causes tending to increase or lower the temperature, should be noticed.

It is only when these several conditions are known that we can hope to arrive at a more exact estimate of the real rate of increase of temperature with depth in Coal Mines. It is to be feared that very few of the observations recorded in Table II satisfy this standard. While they present many points of interest, and confirm the general fact of an increase of temperature with depth, they fail in most cases to give the more precise information required. For these reasons I feel that only a very small selection can be made, and I doubt whether those even give sufficiently true readings, though they may give the best approximation that can at present be obtained.

In making the selection I am aware that I aim at a standard which seems at present not readily attainable: I however give them all rather for future application where deemed fitting. Taking the data as they now stand, we can only select a certain number of observations which best fulfil the required conditions, though we may omit some which deserve attention.

Permanence of the temperature at a station is not, as it has been sometimes considered, always a sign of its correctness, or of its being the true normal. On the contrary, when it is stated that the thermometer in the same hole continues to give the same reading for a long period, it is evident that, instead of having a more definite value, it may represent not the normal temperature of the rock, but that of the rock or coal after cooling to a point at which an equilibrium has been established between the temperature of the strata and that of the circulating air. Such readings therefore will be too low.

Amongst the most careful observations are those at Boldon in the Newcastle coal-field (No. 150). The holes in the rock were there 10 feet deep; the temperature of the air in the gallery was nearly the same as that of the rock; and the mean annual temperature well known.

At the North Seaton Mine (No 217), the station being half a mile distant from the shore, and more than that distance from the shaft, a considerable uniformity of temperature between the rock and the air was ensured.

The experiments at Hetton Colliery (No. 116) seem to fulfil the required conditions, but I do not place implicit reliance on them, because the coal was exposed for at least three days, if not more, to a current of 5,000 cubic feet of cooler air per minute. It is stated, however, that there was no difference in the temperature after the lapse of a fortnight, though this might have arisen owing to the coal having reached a cooling point at which the equilibrium between the air and coal had been temporarily established.

The experiments at South Hetton (No 136) were made in a bore-hole drilled into the strata at the bottom of a shaft, 1066 feet deep, and which was carried to a further depth of 878 ft. The first reading made on the completion of the hole, and while yet dry, gave a temperature at bottom of 96° F. Three years later the experiments were repeated, but in the meantime the hole had become filled with water, and silted up to the depth of 234 feet. The thermometer was, however, pushed down in the silt to the depth of 26 feet, and a temperature of 77° recorded. Whilst the first readings were too high, for the heat caused by the tools could not have been all lost,[1] it is not improbable that convection currents may have cooled the mud in the bore-hole to an approximate mean. The temperature in the bore-hole between the depths of 100 and 670 feet (1166 and 1736 feet below surface), exhibits a difference of 11°, which gives a rate of increase of 52 feet per degree. (This is now replaced by the Whitehaven observations.)

The strong ventilation, and the uncertainty about the length of exposure, are objections to the otherwise careful records of the Pendleton Colliery (No. 104), and the same doubts attach to the other deep works of this district. It is true that many of the observations are said to have been made in newly-opened ground, but we do not know whether this means an hour, a day, or a week, and as in all newly-opened mines the ventilation is well kept up, the rate of cooling must be rapid.

The Wakefield pit (No. 120) had been newly opened. Ventilation was stopped as much as possible at the stations, and the temperature of the outer air was very near the normal mean; the hole in the coal was

[1] The boring was only stopped about twenty minutes before the observations were taken.

6 feet deep, and distant 600 yards from the shaft. In the Barnsley pit (No. 119) the station was too near the shaft, and the air was much below the normal.

The observations both in the Radstock and Kingswood Collieries (Nos. 163 and 155) were made in holes only 2 feet deep, and the thermometers allowed to remain respectively two, five, and seven days—long enough for the rock to cool, as shown by the fact that after another week the thermometer at Kingswood gave the same reading. The temperature of the air is not given.

I pass over the Upper Duffryn, Tredegar, Cwyn Neol, and other collieries, in consequence of the deflections of the isothermal lines by irregularities of the surface.

The particulars of the Norley Colliery, Wigan (No. 111), are insufficient, though the result there obtained seems to confirm the careful observations made at the Rosebridge Colliery.

The observations in the Mons Collieries (Nos. 215, 216) were made at a distance from the shaft, and in dry galleries without ventilation.

At Liège (No. 141) the unusual precaution was taken of making the holes 5 mètres deep, while the temperature of the air in the galleries was only 1° lower than that of the coal.

I have not made use of any of the earlier observations, owing to the uncertainty attaching to the instruments, and, after eliminating others for the reasons named, it seems the safer plan to take only the following small number of observations on which to base our estimate :—

Original number in Table I.	Place.	Depth of pit.	Temperature of rock or coal.	Rate of increase for 1° Fahr.
		feet.		feet.
150	Boldon, Newcastle	1,514	79°	47
226 227	Whitehaven—mean of two pits	1,140 1,250	71 73	49
126	Rosebridge, Wigan	2,445	94	53
217	North Seaton	620	61	45
120	Wakefield	1,455	78	50
141	Liège, Belgium	1,656	87	46
216	Mons, Cuesmes Colliery	1,680	82	54
215	,, La Louvière Colliery	1,456	81	49
			Mean rate of increase	49·1

This seems a very small selection, but I believe that the disturbing causes I have mentioned operate so constantly, that an element of uncertainty is introduced at the very opening of a pit, and that, without the above-named precautions, the observations are not reliable for the purpose of those exact values necessary to determine the true thermometric gradient.[1]

It is only in a few instances that temperature observations have been made by borings in unworked carboniferous strata. In these others there is no indication of the excessive range which the observations in shafts and galleries have shown. The following table gives the results obtained in *bore-holes* sunk in

[1] Instead of making the observations in a newly-opened and well-ventilated part, with fresh faces of the coal emitting pent-up gases, I would suggest, among other means, trying an isolated, abandoned, unventilated gallery, and there place thermometers in holes 5 to 10 feet deep in the coal, and the rocks above and below the coal. The gallery should then be closed at distances of 100 to 150 feet by a series of diaphragms to stop ventilation and convection currents, and the place not opened for several months, by which time the end station would possibly have acquired the temperature of the circumjacent body of strata.

search of coal. At Blythswood the Coal-measures lie under 71 feet of boulder clay, and at Torcy and Mouillonge under 1312 feet and 1217 feet respectively of triassic strata. At the last two pits the beds lie at a considerable angle to the horizon.

Original number.	Place.	Depth.	Rate of increase for 1° F.
		feet.	feet.
123	Blythswood	347	52
124	South Balgray	525	42
68	Creuzot, Torcy Colliery . . .	1,817	57
69	,, Mouillonge Colliery .	2,677	52
		Mean	50·8

§ 4. Observations in Mines other than Coal.
(*Table III.*)

The causes affecting the thermal conditions of Metalliferous Mines are different to those which obtain in Coal Mines. In the former we have to deal almost exclusively with crystalline and schistose rocks. Not only does the conductivity of such rocks differ materially from that of the Coal-measures, but the disturbing causes common to the two have in general very different values.

Ventilation affects them unequally; and while in Coal Mines the influence of water is generally small, it plays an important part in Metallic Mines. On the other hand, chemical decomposition and hot springs, which are common disturbing causes in the latter, are of rare occurrence in the former.

Ventilation.—I have so fully described the effects of ventilation in the section on Coal Mines, that I only need mention here in what respects its effects differ in other mines. Besides, we are not in possession, in respect to metallic mines, of such detailed particulars as those which the Report of the Coal Commission gave of the Coal Mines. The effects are also modified in the two cases by differences in the structure of the shafts and galleries, and while in the one the presence of the coal-gases necessitates excessive ventilation, in the other, hot springs and chemical decomposition, though they raise the temperature, do not render so active a ventilation imperative.

In some of the Cornish mines the current of air is hardly felt, and it is stated generally by Mr. Robert Were Fox,[1] that in deep mines the temperature of the rock and that of the air do not materially differ, except when the currents are strong. At the same time it is evident that strong currents of air do sometimes prevail. An instance is mentioned by Mr. Robert Hunt,[2] where at the bottom of a mine 1950 feet deep, the current of air was so strong that it was difficult to keep a candle alight. In another, in all the levels (which were of great extent) of one hot lode at the depth of 1410 feet, the ventilation was so effective that the temperature of the air was never higher than 70° F.

These mines must suffer, as in coal mines, a great and durable reduction of temperature in the proximity of the down shafts. This is especially noticeable in

---

[1] *Coal Commission Report*, vol. ii., p. 211.
[2] *Ibid.*, vol. ii., p. 87.

severe winters, and when the ventilation is active the effects of this may extend to a considerable distance from the shaft. Daubuisson mentions that during a severe winter, and with the outer air at a temperature of $-15°$ R. ($-2°$ F.), the shaft of the Beschertgluck mine, in the Freiberg district, was lined with ice to a depth of 80 toises (480 feet), and that the temperature of the air at the bottom of the shaft was $\frac{1}{2}°$ R. (33° F.). Dr. Clark ("Travels in Scandinavia") says that in descending the great iron mines of Persberg, in Sweden, which are 450 feet deep, he found large masses of ice covering the sides of the walls, and that it had accumulated in large quantities in the lower chambers. But the formation of ice is not confined to these more rigorous winter climates, for Mr. Moyle speaks of finding, on one occasion, ice in abundance in the Tin Croft Mine, Cornwall, at a depth of 318 feet below the surface, and says that the ladders became impassable, crevices were filled, and icicles formed all around! That this must produce a lasting effect is clear from the circumstance observable in some coal mines, where the walls of deep pits near the shaft are so cooled in winter, that in summer the air circulating by them is of a higher temperature than the rock. On the other hand, the temperature of the air in a metallic mine is more apt than in a coal mine to become heated by the combustion of candles, the explosion of gunpowder, and the presence of the workmen.

Some of the observations of Mr. Fox show the effects of ventilation in the Cornish mines. Thus at Huel Damsel Mine (No. 21), the temperature of the air in the galleries of different levels varied very little, and was higher in the third level than in one lower.

Depth.	Temp. of air.	Depth.	Temp. of air.
480 to 540 feet	69°	780 to 840 feet	73°
600 „ 780 „	70	840 „ 900 „	70

The influence of the air upon the temperature of the rocks is shown by another series of experiments made by him in holes in the rock in Dolcoath Mine with the following results:—

	Depths.	Temperature of rock.	Rate of increase per 1° F.
1.	240 to 300 feet	58°	34 feet
2.	540 „ 600 „	58	71 „
3.	720 „ 780 „	63	57 „
4.	1140 „ 1200 „	64	83 „
5.	1320 „ 1380 „	78	48 „

Mr. Fox does not give the temperature of the air, but states that the abnormally low temperatures of Stations 2 and 4 arose from the passage of strong currents of air. The effect of these conditions, in estimating the rate of increase of temperature with depth, is clearly shown in the last column of the above list.

But although there may be extreme cases, it is probable, as a general rule, that the ventilation does not produce the extent of difference between the temperature of the air and of the rock that it does in Coal Mines. Mr. Fox, who, to avoid the effects of ventilation, always, if possible, made his observations near the ends of the levels, states that in those cases there is little difference between the temperature of the air and the rock. At the same time it is possible that, even then, the uniformity may be owing to the rock having permanently cooled down to the temperature of the air—though it may not be much. The following are cases in which the temperatures of both are given :—

	Depth.	Temperature.		Rate of increase per 1° F.
		Air.	Rock.	
No. 71. Par Consols	1248 feet	82°	84°	38 feet
No. 72. Botallock	1128 ,,	81	79	40 ,,
No. 77. Levant	1530 ,,	85	85	45 ,,
No. 79. Tresavean	2112 ,,	91·5	90·5	52 ,,

Here it will be observed that there is a difference of 1° to 2° between the temperature of the air and the rock, and that the rate of increase with depth is, with one exception, much more uniform.

May not circumstances such as these account for the marked discrepancies in the rate of temperature with depth observed in the Talargoch Mine (No. 158, 160); for, although the temperature of the air is not there given, we know that it will vary with the distance from the shaft :—

Station.	Depth.	Distance from shaft.		Temp. of rock.	Rate of increase with depth.
3	660 feet	120 yards	S.	54° F.	132 feet
5	555 ,,	170 ,,	S.E.	52.9	140 ,,
4	465 ,,	190 ,,	S.W.	53.4	116 ,,
1	1041 ,,	190 ,,	N.E.	60·8	88 ,,
6	636 ,,	840 ,,	S.W.	58·8	64 ,,
7	660 ,,	1240 ,,	S.S.W.	62	51 ,,

The two partial exceptions (4, 5) are in both cases at stations of lesser depths than the others, and, being near the shaft, were possibly more exposed to the influence of the outer air than the others.

Some of the early experiments of Dr. Forbes (No.

18) were made in the water in *sumps* and pools, which necessarily would be much exposed to the cooling effects of the circulating air currents. These observations are also at variance with those subsequently made by Mr. Fox himself and by Mr. Henwood. The following are some of the thermometric gradients obtained from these experiments at various depths :—

Depth.	Temperature of the water.	Rate of increase per 1° F.
500 to 550 feet	65°	**35 feet** [1]
900 ,, 950 ,,	71	**44** ,,
1,350 ,, 1,400 ,,	79	**47½** ,,

Other observations of Mr. Fox, which were made in the same way in the water in fifty-three mines, gave also a diminished rate of temperature with increase of depth :—

Mean depth.	Rate of increase. per 1° F.
354 feet	**35·4 feet**
448 ,,	**43·8** ,,
648 ,,	**64·2** ,,

Mr. Fox's later observations in rock give different and apparently more reliable results.

The Effects of the Percolation of Water.—But while the effects of ventilation are not so general and disturbing in Metallic Mines as in Coal Mines, the effects produced by the underground waters are of greater importance. The alternation of impermeable

---

[1] The rapid rate constantly observed near the surface is probably in great part owing to chemical decomposition in the lodes which is most active near the surface, and also to a less active ventilation.

with permeable strata, and the multiplicity of faults in the Coal-measures, so impede the descent of the surface-waters, that there are mines so dry as to necessitate the introduction of water to keep down the dust. The Metallic Mines likewise, which are commonly in crystalline, schistose, and slaty rocks, are generally more or less impervious. When, however, they have been disturbed and fissured, they give free passage to water; and when, further, they are traversed by veins and faults, these frequently serve as channels or conduits, more or less free, for the surface-waters, and considerable quantities of water pass through them. Consequently water, which finds its way to depths with more or less rapidity, is often one of the great obstacles to deep mining with which the workmen have to contend. Mr. Henwood states that in some mines a great increase follows soon on heavy autumnal rains, and that in others, months intervene before the effects are felt.

In districts formed of the usual alternations of sedimentary strata, it is estimated that on an average about one third of the rainfall passes underground; while in Cornwall, where granites and slates exclusively prevail, Mr. Henwood estimated in his survey of the Gwennap district—which consists chiefly of slates—that about a fourth of the rainfall passes underground, the mean annual rainfall there being 46 inches, or equal to 166,834 cubic feet per acre. The local percolation is, however, extremely variable, as in some mines the quantity pumped up does not exceed 5 gallons, while in others it amounts to 186 gallons per minute.

The same observer found that water passes more freely through Slate than through Granite, the quantity yielded by mines in slate being about four times as much as that in granite. In a period of five years (1833–7) the water pumped per minute from forty mines amounted on a mean to the following proportions:—

Granite.		Slate.		
Maximum.	Minimum.	Maximum.	Minimum.	
30	16	122	63	cubic feet per minute.

Mr. Henwood's account of the quantity of water that passes underground in the Gwennap district is of much interest. The mines in that district, over an area of 5,500 acres, were combined for drainage purposes. One great adit carried away the water standing above the sea-level, while another deep-seated level collected the water at the depth of from 1,100 to 1,200 feet. At the time that Mr. Henwood wrote, the former was discharging 1,475 cubic feet per minute, and from the latter 909 cubic feet were being pumped up, or together above 10,000,000 gallons in the twenty-four hours. Taking the mean temperature of the surface at 50°, as the water issues at a temperature of from 60° to 68° F., or at an average of more than 12° above the mean of the climate, it is easy to conceive how large must be the amount of heat which the waters abstract from the mines, and how considerable the cooling of the adjacent rocks which must result therefrom.

Another observer, writing a few years earlier, states that the discharge at the Huel Vor Mine from a depth of 950 feet was 1,692,660 gallons every twenty-four hours; at Dolcoath Mine, from about 1,400 feet, 535,173 gallons; and at Huel Abraham Mine it reached the large quantity of 2,098,320 gallons.[1]

Mr. Henwood remarks that "the largest streams of water flow through cross veins; small ones through the lodes, whilst but little issues from the rocks whether granitic or slaty."

Where the water dribbles slowly through the rocks to great depths, it will probably acquire the normal temperature of the depth, but where it passes more rapidly through the veins and lodes, the temperature will depend upon the time occupied in transit and on the volume of water. If the flow is rapid, as it evidently is in some mines, the surface-waters may carry the influence of the above-ground temperature to considerable depths. If, on the other hand, the vein is one in which the ore is subject to decomposition by the surface-waters, those waters will have their temperature more or less raised. A copious stream of warm water is considered among the Cornish miners a favourable indication of the proximity of a lode. Nevertheless, Mr. Henwood considered that by a careful selection of the underground springs and by taking them when freshly opened, they gave safer temperature results than did the rock.

HOT SPRINGS.—These are not uncommon in metallic mines. They are due to two causes:

[1] Dr. Forbes, *Trans. Roy. Geol. Soc. Cornwall*, vol. ii., p. 167.

1st, to chemical decomposition; 2nd, to water rising from greater depths.

The first of these causes is a very general one—especially in those mines where the lode consists of iron and copper pyrites. The surface-waters decompose these sulphides, converting them into sulphates, which by further changes pass into the oxides and carbonates of these metals. That the action is general is shown by the circumstance that the upper part of all these lodes consists near the surface of a crust, several feet and sometimes several fathoms thick, composed of the oxidised products of copper and iron sulphides; this part of the vein is known in Cornwall under the distinctive term of *gossan*. In these cases, the water is commonly impregnated with some of the resulting soluble sulphates, and has its temperature raised by their decomposition.

Mr. R. Hunt mentions [1] two marked instances of the heating effects arising from this cause. In one case the temperature in the level of a copper mine stood at 100° F., but on the removal of a very large deposit of the copper pyrites "the level became cold enough to make the agent wish for a great-coat." The exact difference is not given. In another case, a large deposit of iron pyrites was opened at about half a mile distant from a hot lode, 1,530 feet deep, in the Clifford Amalgamated Mine, and the mere fact of opening the mine there and removing the iron pyrites considerably reduced the temperature in the other mine. When it was closed the temperature rose

[1] Coal Commission Report, A 4-10.

to its former height. Springs of various degrees of heat (one was as high as 124°) are often met with.

The miners of Cornwall have long observed that the lodes containing tin are, at equal depths, colder than those in which copper ores occur; a fact which is no doubt due to the facility with which the cupreous pyrites decompose.

There must also be cases in which water from greater depths is brought up along lines of fissure (lodes and cross veins); for as these are prolonged downwards, they may, at greater depths, traverse strata, veins, or faults, charged with water, which, when thus tapped, will outflow in the manner of an artesian well.

It sometimes also happens that at the same depths, but in distant parts of the same level, the water is of different temperatures. In one instance there were springs respectively at 102°, 110°, and 124° Fahr., and in another case (Wheal Wreath), the temperature of a small stream at the east end of a lode (tin) was 71·5°, while a spring at the west end of the lode had a temperature of 75° (both being at the depth of 1,422 feet).

On the other hand, when the water is in considerable volume, and percolates rapidly, it must tend to have a lower temperature than the normal rock temperature, as in the instance where, in two adjacent mines, large streams, both coming out of veins, had the same temperature of 67·5° at the respective depths of 588 and 722 feet. The observations of Henwood likewise tend to show that the range of temperature of water in the same level is subject to great variation.

In consequence of the uncertainty attaching on the one hand to observations taken in rock, and on the other to those in water, Cornish geologists have been divided in opinion as to the best plan to adopt. The two great authorities on the subject, Mr. R. Were Fox and Mr. W. J. Henwood, respectively gave preference —the one to rock and the other to water (or rather springs). Either system is in fact open to some objections, and neither can be employed without recourse to those safeguards and precautions which no one knew better how to use than these two experienced observers.

Mr. Henwood gives as his reason for preferring springs—that the rocks forming the sides of the shaft and levels must, to a certain extent, partake of the temperature of the air circulating through them, and that this air could not escape the influence of heat-producing causes at one time, and at another of the cooling effects of the intake air. He objected, however, to the use for temperature purposes of the water standing in the levels, or *sumps*, and he was led to confine his observations as much as possible to the temperature of the streams of water immediately as they issue from fresh opened unbroken rock—before they could be affected by the temperature of the levels—as the places which would give the most correct temperature readings.

The following observations of Mr. Henwood appear amongst the most reliable of those obtained in springs. The deepest seated springs seem the most free from interference by the surface temperature and other disturbing causes :—

No. in Table.	Mines.	Depth.	Temp.	Therm. gradient.	Spring in rock or lode.
		feet.	Fahr.	feet.	
52	St. Ives Consols	810	71°	41	Rock.
49f.	Wheal Trenwith	660	66	44	Lode.
49g.	South Roskear	834	71	40	Lode.
49a.	East Wheal Crofty	810	70·7	39	Rock.
79	Tresavean	2,112	93·5	48·5	Rock.
27	Dolcoath	1,440	82	45	Lode.
49b.	Consolidated	1,764	92·5	41·5	Lode.
49c.	United Mines	1,080	74	45	Lode.
49l.	East Wheal Virgin	1,722	94·5	39	Rock.
51	Devonshire Wheal Friendship	810	69·5	41·5	Lode.
			Mean	42·4	

In another instance Mr. Henwood, taking the mean of 415 observations made in ten Cornish mines, obtained the following rate of increase of temperature with depth :—

	Average depth.	Mean temp. at depth.	Rate of increase per 1° Fahr.
1.	180 feet	54·8°	36 feet
2.	432 ,,	60·8	40 ,,
3.	762 ,,	67·4	43·5 ,,
4.	1,050 ,,	78·0	40 ,,

He elsewhere gives the mean temperature for other depths :—

Mean depth	Mean temp.	Rate of increase.
672 feet	66·9°	40 feet
1,440 ,,	85·5	40·5 ,,

He noticed also that there was a difference between the temperature of the Springs issuing from the granite and those from the slate rocks. The mean of 134 observations made to a depth of 1,200 feet in

Cornwall and Devon, gave the following "thermometric gradients":—

In Granite	41·5 feet per 1° Fahr.
In Slate	39·0 ,, ,,

Further, taking separately the springs issuing from veins of different characters, he arrived at the following results:—

For cross veins at a depth of	594 feet......	40·0	feet per 1°
For lodes generally ,,	660 ,, ...	41·6	,,
For Copper lodes ,,	840 ,, ...	38·0	,,
For Tin lodes ,,	552 ,, ...	51·5 [1]	,,

The result of these and various other calculations is that he obtained from the Springs issuing from the Rock a mean "thermometric gradient" of 40·1 feet, and from the Lodes of 40·3 feet per degree.

The early observations of Mr. Fox were made in the mine Waters, but his later experiments were made in the Rocks, and he expressed afterwards some doubts as to the value of his earlier experiments.[2] He considered that the rock observations were free from the direct influence which the descent of the surface-waters exercises on the lodes and cross veins—of which the effects may either be to raise the temperature of the underground springs when chemical decomposition is going on in the upper part of the lode, or

[1] There would appear to be some mistake either in this or in a figure in another page, where he states that the temperature in the tin and copper lodes conjointly at a mean depth of 444 feet is 61·4°, which gives a rate of increase of 39 feet per degree.

[2] Some were made in still water or *sumps*; others in the air of the levels.

to lower the temperature when the surface-waters are abundant and percolate rapidly.

The following observations made in the rock appear to be amongst the most reliable of the results obtained. The temperature was taken in holes 2 to 3 feet deep, and the stations were selected at places where no working had recently been going on, and as distant as possible from the shaft:—

No. in Table I.	Name of mine.	Rock.	Depth.	Temp. of rock.	Rate of increase for 1° F.	Observer.
			feet.		feet.	
19	Dolcoath	Granite	1,350	78°	48	Fox.
76a–77	Levant	Granite and slate	1,530 1,530	87 85	43·7	,,
72	Botallock	Granite and greenstone	1,128	79	40	,,
80	Fowey Consolidated	Slate	1,728	93	41	,,
71	Par Consols.	Slate	1,248	84	38	,,
40	Consolidated	Slate	1,740	85·3	49	Henwood.
122a	Tresavean	Granite	2,130	99	43·5	Hunt.
78	Tresavean	Granite	1,572	82·5	48·5	Fox.
					Mean . 44 feet	

This gives a mean thermometric gradient of 44 feet per 1° F., or combining the results obtained by observations in springs with those in rocks, we get an average gradient of 43·2 feet per degree.

```
Mean of observations in Springs .........  42·4 feet
      ,,              ,,       Rock ............  44·0 ,,
                                              ─────────
                                      Mean...  43·2 ,,
```

Foreign Mines.—We know too little of all the conditions which obtain in foreign mines to draw any definite conclusion as to the rate of increase of temperature. Possibly the observations may prove available when we have more certain information respecting the mean annual temperature at each place, the height of the mine above the sea, and the position, especially in mountainous districts, of each station with reference to its depth beneath the surface.

This latter element is one which the coal-mines observations previously given clearly show should be taken into account. In Cornwall it is not of much importance, as the elevations are small, and the mines are rarely more than 100 to 300 feet above the sea-level. The mines of Freiberg are on much higher ground, whilst those of Prizbam and Chemnitz are situated amongst high hills, and the temperature at the end of a long gallery may be, in relation to depth beneath the surface, very different to that given by the depth of the shaft.

The mines of Freiberg are those in which the greatest number of observations have been made, though none of them are of very recent date. The deepest mine, and one in which the observations were made in holes in the rock (gneiss), gives, if we are right in our estimate of the surface temperature,[1] a rate of increase with depth of 54 feet per degree. But in this instance (No. 13)—as in the case of Mr. Fox's observation (No. 73) in Dolcoath Mine, where the thermometer was left in position for one and a

[1] This is based on the temperature of Dresden, the nearest place where we have recorded observations. The mean annual temperature of Dresden is 47°, and as the mine is situated at the height of about 1,300 feet, or about 900 feet above Dresden, if we allow 1° for every 300 feet of elevation, we shall have about 44° for the surface temperature at the mine.

half year—it was here left two years, and in all probability there was, as before described, a cooling of the rock, such as would reduce the temperature so far below the normal, as to place it in equilibrium with the temperature of the air in the galleries. The rock temperatures of 66° at Freiberg, and of 76° at Dolcoath, with their thermometric gradients of 54 and 53 feet, may therefore represent the temperature of the rock, minus the loss of heat due to ventilation. If that can be estimated at anything like the loss shown to have taken place in a level in the Wheal Vor Mine, which,—after it had been opened some time and the mine had been deepened—amounted to about 6°, or in a similar instance mentioned by Mr. Henwood, where the difference amounted to 7°, those figures would indicate, after allowing for that loss, a normal temperature of both these mines more in accordance with the gradient of 44° which we have adopted for Cornwall.

For the same reason the numerous observations made by Mr. Henwood in the mines of Brazil and Chili, are, for the present, not available, though it is possible that they may be rendered so at a future period, should the other factors necessary for determining the difference between the surface and the underground temperature be ascertained.

§ 5. Artesian Wells and Borings.
(Table IV.)

This class of observations presents results much more uniform than those taken either in Coal or Mineral Mines, and whereas the observations in Mines were in Palæozoic or crystalline rocks, those in Wells are, with few exceptions, either in secondary or Tertiary strata, where the rocks are, as a rule, more permeable and less disturbed.

Many of the interfering causes, difficult to eliminate in the observations on the Mines, do not exist with the Artesian Wells. The causes of interference in the latter are reduced mainly to two—namely, pressure on the instruments and convection currents. The early experiments, where precautions were rarely taken against pressure, are consequently unreliable. Walferdin introduced improved and protected instruments, and in some other previous cases, as in Marcet and De la Rive's observations, protected thermometers had been used. The need of protection against convection currents had also not escaped attention, but it was not until the later observations instituted by the Committee of the British Association were made, that more efficient safeguards were introduced to protect against the subtle influence of these currents.

It is clear that we must reject all the early experiments made with unprotected thermometers; and it is not certain whether also a large number of those made with protected thermometers, but without protection, or sufficient protection, against convection currents, should not also be rejected. In large bore-holes the disturbance from this cause is so great that the consequences are at once sufficiently apparent to warrant the rejection of the observations.

The deep Artesian wells of Paris, such as Grenelle and other similar ones, agree in showing a rate of increase of 50 to 55 feet per degree F.; but the great bore-hole ($4\frac{1}{4}$ feet in diameter) through partly the same strata in another quarter of Paris (La Chapelle St. Denis), where the water does not overflow, gave a rate of increase of $39\frac{1}{2}$ feet per 1° for the first 100 mètres (328 feet), which is too rapid, whilst at the depth of 660 mètres (2,165 feet) the rate of increase was only 1° in 84 feet. This is clearly due, as stated by Professor Evrett, to the overheating of the upper part of the column of water and to the cooling of the lower part by the action of convection currents. Of a deep boring at Moscow it is stated, but without explanation, that from 350 feet to the bottom at 994 feet the temperature was nearly constant at 10·1° C. (50·2° F.). The diameter of the bore-hole is not given, but I judge from various circumstances that it was not less than 2 feet, and can only account for the uniformity of temperature by the action of convection currents. The mean annual temperature of Moscow is 39·4°.

Professor Evrett[1] also directs attention to the manifest action of convection currents in a shaft at Allendale, about 350 feet deep, with nearly 150 feet of water, in which the temperature was practically the same at all depths.

[1] *Brit. Assoc. Reports* for 1871 and 1869.

Depth.		Temp.
160 feet	....	47·5°
200 ,,	....	47
250 ,,	....	47·7
300 ,,	....	47·7

One of the pits at Allenheads (No. 130) which was nearly full of water gave similar results.

Depth.		Temp.
50 feet	....	47·2°
100 ,,	....	46·8
200 ,,	....	46·6
350 ,,	....	46·9

While in another shaft at Ashburton 620 feet deep, and with water standing to within 50 feet from the surface, the temperature at all depths was 53°, except at one point, where it rose to 53·4°.

Even in the deep and narrow bore-hole at Sperenberg, it was shown that the first experiments with thermometers protected against pressure, but not against convection currents, gave wrong results, the temperature in the first instance, at a depth of 100 feet, recording 11° R., and at 3,390 feet, 34·1° R. Afterwards, when plugs were inserted to stop the currents, the temperature at the same respective depths was found to be 9° R. and 36·16° R., showing that the first readings were too high by 2° Réaumur near the top of the bore, and too low by 2·05° Réaumur at the bottom. Later experiments in the same well showed that the difference at the bottom, between plugging and no plugging, was $6\frac{3}{4}$° F.[1]

Another bore-hole, showing clearly the action of

[1] *Brit. Assoc. Report* for 1876, p. 205, and 1882, p. 3. The differences were even considered to be under-estimated.

convection currents, was the one sunk to a depth of 2,000 feet at Swinderby, Lincoln (No 143). The hole had remained undisturbed for nearly three weeks, and the water stood within a few feet from the top.

Depth.	Temp.
100 feet	68° F.
300 ,,	68·75
500 ,,	68·75
600 ,,	69
900 ,,	69
1,200 ,,	69·5
2,000 ,,	79

Taking the mean annual temperature of Swinderby at 48°, this would give a difference of 31° for the whole depth, or $64\frac{1}{2}$ feet increase for each degree F. If it stood by itself this might appear nothing remarkable, yet it is evident, from the series of observations at different depths, that the rate of increase at the depth of 100 feet, of 1° in 10 feet, and at 300 feet, of 1° in 28 feet, is excessive, and that this excess can only have been acquired through loss of heat by convection currents at the bottom and a corresponding gain at the top, and thus making the readings too high at top and too low at bottom. If, however, we take an intermediate station at a depth of 1,000 feet, where the temperature averages 69·2°, we get a mean, and more probably approximate, rate of increase with depth of 47 feet per degree.

The great well at Bootle (Liverpool) gave very unreliable results owing to the large diameter of the borehole, the many water-levels in the New Red Sandstone, and the strong convection currents (No. 153).

It is clear, therefore, that great uncertainty attaches to all observations made in bore-holes with standing water, the error being in proportion to the diameter of the bore-hole; and that where experiments have been made without plugging, all the deep temperature readings will be misleading. Even with this precaution, it may be a question whether the bottom water and that of the adjacent rock may not have had their temperature permanently lowered before trial.

There are, however, some artesian bore-holes where the sources of error have been reduced to a minimum. Among those are—

1. *Kentish Town* (No. 129).—Careful experiments were carried on there for some years by Mr. G. Symons, F.R.S. From the circumstance that the mud at the bottom of the bore-hole into which the thermometer was sunk was little affected by convection currents, the results obtained show, possibly, a near approximation to the normal temperature of the rock. At the same time the long period that the well had stood neglected allowed the play of those currents in the water standing in the tube above the mud, and this may probably have effected a slight reduction of temperature.

2. *Richmond* (No. 230).—The observations here were made by Professor Judd with standard instruments. The overflow water was in too small a quantity to give the correct temperature, though enough to check convection currents. The temperature was ascertained by letting down a thermometer to the bottom of the well.

3. The first observations at *Grenelle* (No. 37) were made when the bore-hole had reached the depth of 400 mètres (1,312 feet), and when no work had been going on for some weeks. The thermometers, of which three sets were used, were left down thirty-six hours. They were protected against pressure, and Arago remarks that the chalk through which they were boring made so thick a paste filling the bore-hole, that convection currents were hardly possible. All three sets of instruments gave results within a fraction of a degree to one another.

4. In the bore-hole of the well at the *École Militaire* (No. 36) the experiments were made by Walferdin under very similar conditions.

5. *Pregny near Geneva* (No. 29).—The thermometer was protected against pressure, and to a certain extent against convection currents. The bore-hole was small, and the water stood at a small depth below the surface.

6. The observations at *Moullonge* and *Torcy* (No. 68) were made by Walferdin, who, to secure accuracy, employed eighteen protected thermometers.

7. At *Sperenberg* (No. 144) especial care was taken against convection currents, though it is possible that these currents may have, in some degree, influenced the result. This is a risk to which all non-overflowing artesian wells are liable.

The preceding sets of observations give a mean rate of increase of 52·2 feet for each degree F. (p. 228). For various reasons I have not included a larger number, such, for example, as, amongst many others, those described at the following localities.

Swinderby.—I have already explained my objections in this case. Of the first series of observations down to a depth of 1,500 feet, Professor Everett remarks "it is obvious that nearly all the temperatures are largely affected by convection," he considered the bottom temperature at 2,000 feet as less likely to be vitiated by convection in consequence of the small diameter of the bore-hole.

But if we take, as I have suggested, a mean depth and a mean temperature, I believe we might have in the thermometric gradient of 47 feet per degree, a nearer approach to the true normal at Swinderby. Even in cases where the temperature is uniform from top to bottom, as in the instance with the Moscow Well (No. 135), where Professor Lubinoff records a temperature of 10·1° C. (50·2° F.), if not for the whole depth of 994 feet, at least from 350 to 994 feet, it seems possible to obtain an approximate gradient. For the mean annual surface temperature being 39·5° F., and the half depth 494 feet, if we divide this by 10·7° (the difference between the surface and the well temperature), we get a quotient of 46·5, which agrees nearly with the thermometric gradient of the well at St. Petersburg (No 44).

Southampton.—The well was too long disused, and there was apparently no protection against convection currents.

Rouen, St. Sever.—The thermometer was not protected against pressure, though the effect of the pressure was estimated and allowed for.

Troyes.—M. Walferdin thought it probable that

the observations were affected by the heat caused by the boring tools—sufficient time not having been allowed to elapse after working before the thermometer was sent down.

Artesian Wells and Bore-holes in which the Water stood below the Surface, or rose so slowly that the temperature observations had to be taken in the Bore-hole.

Number in Table I.	Place.	Strata passed through.	Depth.	Temperature.		Thermometric gradient.
				At depth.	Mean on surface.	
			feet.	Fahr.	Fahr	feet.
129	Kentish Town	Tertiary, Chalk, and Old Red Sandstone	1,100	69·9°	49°	52·3
231	Richmond	Tertiary, Chalk, Jurassic, &c.	1,337	75·5	49·6	51·5
37	Grenelle, Paris in the chalk	Tertiary, Chalk, and Greensand	1,312	74·7	51	55
36	École Militaire, Paris	Tertiary and Chalk.	568	61·5	51	54
29	Pregny	Tertiary (Molasse).	713	62·7	47	48·5
69	Creuzot, Moullonge.	Triassic Sandstone on Coal Measures.	2,677	100	48·5	52
144	Sperenberg	Triassic Rock Salt and Gypsum	E. ft. 3,490	115·5	48·3	52
					Mean	52·2

Overflowing Artesian Wells.—We now come to surer ground. Under certain conditions, these wells must give not only the nearest, but a very near, measure of the underground temperature, at the level from which the water rises. These conditions are—

1st. A sufficient depth and a sufficient distance of the outcrop of the water-bearing strata from the point of overflow. In a case like the Grenelle well, which is nearly 2,000 feet deep, and where the stratum which serves as a channel for the water does not come to the surface for a distance of above 100 miles, these conditions are most favourable. It is very interesting to find, from information which M. Daubrée has just (Dec., 1884) obligingly obtained for me, that the temperature of the water at the Grenelle well is now the same (27·8° C.), if not a little higher, than it was at the commencement of the overflow forty-two years ago. In London the distance of the wells from the outcrop of the Chalk being only a few miles, it is a question whether the excessive pumping and constant passage of water may not have cooled the channels and lowered the normal temperature of the water in those wells.

2nd. A sufficient volume and rapid upward flow of the water in proportion to the depth, otherwise the water will part with some of its heat, as it rises through the tube. This may possibly be in some small degree the case at Grenelle, for in the carefully conducted observations of Arago and Walferdin, when the well had reached a depth of 400 mètres, the temperature (23·75° C.) gave a rate of increase of 55 feet per degree, instead of that of 58 feet when it first overflowed.[1]

[1] Whether the tubing affects the results I am unable to say. In deep borings the dimensions of the tubes vary very considerably, and may have some influence on the temperature of the water. If the tubes were of the same diameter throughout, the passage of any given portion of the water from the depth to the surface would be direct, and the velocity the same throughout; but the tubes in these deep wells decrease in diameter from the top to the bottom, sometimes considerably; consequently, instead of the whole body of water being in continuous and uniform motion throughout the length of the tube, the velocity of the

In the case of the great saline wells in Germany, especially in the instance of the well at Neu Salzwerk, the conditions of depth and volume are generally favourable for observation. The outflow was, at the time of trial, at the rate of 422,000 gallons daily. It was accompanied by an enormous discharge of carbonic acid. The deep artesian wells of Minden, which derive their water from the same source, gave very similar results.

The discharge of water at the thermal springs of Mondorff in Luxembourg is much less than at those of Neu Salzwerk, and uncertainty attaches to the height of the ground and to the mean annual surface temperature.

The depth of the artesian well at Tours is much less than that of the above, but the ascent of water was rapid and the discharge large. The observations there were made by Walferdin, as were also those at Rochefort.

The following list is confined to those wells where the overflow is abundant, and where the observations have been made by competent observers: —

water gradually decreases from the bottom to the top, currents are established at the points where the tubes enlarge, and the direct trajectory is delayed. For example, at Grenelle there are four tubes of the respective diameters of 0·25 m., 0·22 m., 0·18 m., and 0·17 m., with a difference between the bottom and top of $3\frac{1}{4}$ inches in the diameter of the bore-holes. This may possibly help to explain the reason why the wells of 300 to 600 feet deep, where there are fewer changes of tubes, give generally a more rapid rate of increase than the wells of 2,000 feet, and may also account for the different rates of increase at the several wells for which otherwise there is no apparent cause.

Artesian Wells in which there is an overflow of Water at the Surface.

Number in Table I.	Place.	Strata passed through.	Depth.	Temperature.		Rate of increase per 1° Fahr.
				At depth.	Mean on surface.	
			feet.	Fahr.	Fahr.	feet.
48a	Paris, St. Ouen	*Tertiary*	216	55·3°	51°	50
37	Paris, Grenelle.	*In Greensand*	1,797	81·9	51	56
32	Lille	*Cretaceous ; Carboniferous limestone*	329	57·2	50·5	49
31	Tours	*Cretaceous*	460	63·5	53	44
101	Rochefort	*Trias*	2,812	111	54·5	50
65	Mondorff	*Lias and Trias*	1,647	78·3	47·3	53
182	Minden	*Triassic (?)*	2,230	90·9	48	52
34	Neu Salzwerk	*Liassic and Triassic*	2,038	88·3	48	50·5
					Mean	50·6

The thermometric gradient of this group is 1°·6 less than that of the first group; the mean gradient for the two being 51°·4.

I have excluded many wells because of the uncertainty attaching to the instruments used, or to the omission of some essential particulars. These reasons apply to such wells as those of Newport (No. 210), Falkirk and Midlothian (45), Dunkirk and Bourbourg (209, 208), Alfort (49), Meaux (62), Arcachon (183), and others, where we do not know whether or not standard and protected instruments were used, or whether the experiments were in all such cases made under the right conditions.

With respect to the extra-European observations, still greater uncertainty attaches from our ignorance of the local conditions, and especially of the exact mean annual temperature of the several places. At the same time there are some exceptions worthy of consideration. The experiments in the Sahara Desert

(No. 88a) were made by an engineer of great experience in the construction of artesian wells and accustomed to observations of this description. The mean is from observations taken at a number of wells. The observations at Charleston and St. Louis appear reliable, and are interesting from their great depth, but fuller data are needed.

The African and Indian experiments seem to indicate a more rapid rate of increase of temperature with depth than occurs in Europe. Not so the American (U.S.) observations, which appear to indicate conditions very similar to those which obtain here. Not much weight can be attached to the solitary observation in South America.

§ 6. Tunnels

The few observations of this class, limited as they are, show not only the modifications of the gradients caused by inequalities of surface, but bear also on some important geological questions connected with the structure of mountain chains and metamorphism. The first great tunnel was that of Mont Cenis. It is about seven miles long, and passes under the axis of the Alps which is there 9,532 feet above the sea-level, and 5,280 feet above the tunnel. After making a correction for the convexity of the surface, Prof. Everett estimates the rate of increase of temperature with depth to be 1° F. in 79 feet. But the tunnel was far advanced before the observations were commenced.

In the St. Gothard tunnel, the observations, which were carried out by Dr. Stapff, were more continuous and complete. The tunnel is about nine miles long;

the summit level of the ridge above the tunnel is
10,040 feet above the sea-level, and 5,578 feet above
the tunnel. This, after allowing for the convexity
of the surface, gave a rate of increase of 1° F. in 82
feet. But Dr. Stapff[1] has since pointed out that in
one part of the tunnel the rate is considerably more
rapid. He found that the relative temperature of
the ground above the northern end of the tunnel was
much higher than in other parts—that in the plain
of Andermatt the mean rock temperature was several
degrees above the normal, while at the south end of
the tunnel it was some degrees below it. The latter
circumstance was easily explained by the presence of
cold springs. Some higher temperatures in other
parts of the tunnel were attributable to the decomposition of the rock; but there were no apparent reasons
for this excess of temperature in the northern end of
the tunnel, where it passes through gneiss and granite.
The difference was such that instead of the rate of
increase of 1° in 85 feet, as in the centre of the tunnel,
or of an average rate of increase for the whole tunnel
estimated by him at 57·8 feet, the rate was here 1° F.
in 38 feet. Dr. Stapff says that there is no obvious
explanation of the rapid increase in the granite rocks
at this end of the tunnel, and that it is probably to
be attributed to the influence of different thermal
qualities of the rock. He mentions, further, that this
granite belongs to the *massif* of the Finsteraarhorn,
which is of a different (newer) geological age to that
of the central axis of St. Gothard, and that it is

[1] *Trans. North of England Inst. Min. and Mechan. Engineers*, vol. xxxiii. (1883), p. 19.

therefore not to be wondered at "if one of them be cooler than the other." He elsewhere remarks that there is also a well-known local focus of heat (decomposition of rock) below the valley of Andermatt, which may exercise some influence.

I myself am more disposed to attribute the greater heat of these rocks to mechanical action rather than to the later protrusion of the Plutonic rocks. If the pressure and friction accompanying elevation of mountain chains be attended by the development of heat—a heat sufficient to produce great chemical changes even in the Tertiary strata—then it may be possible for some of the newer mountain chains still to retain a portion of the heat thus developed. The facts brought forward by Dr. Stapff in the St. Gothard tunnel give material support to this view.

Although Mallet failed to show that the heat produced in the crushing of rocks by the lateral pressure, arising from the contraction of the crust of the earth as a consequence of its slow secular refrigeration, was sufficient to fuse the rocks and account for volcanic phenomena, he nevertheless brought prominently forward the enormous heat-producing power that might result from this cause.[1] He made a series of elaborate experiments to ascertain the force required to crush blocks of a given size (about 1.5 inch on the edge), and measured the work done by the *estimated* heat evolved by the crushing of 1 cubic foot of several classes of rock by the number of cubic feet of water at 32° F. converted into steam of one atmosphere, or

---

[1] On a small scale this is clearly shown by the *creeps* and crushes in coal-pits (*ante*, p. 192).

at 212° F. This method, although not perfectly satisfactory, is sufficient to prove the essential fact that a mechanical disturbance of the rocks may develop a large amount of heat.[1] The following are abstracts from his table of experiments:—

Class of rock.	Specific gravity. Water = 1,000.	Weight (pressure) per square inch at first yielding.	Mean pressure at which the cubes were completely crushed.	Temperature of 1 cubic foot of rock due to work of crushing.	Number of pounds of water at 32°, evaporated into steam at 212°.
		lbs.	lbs.	Fahr.[2]	lbs.
Caen Oolite	2·337	1,620	4,966	8°	0·288
Magnesian Limestone .	2·571	3,699	16,333	26	0·9
Coal-measure Sandstone.	2·478	10,970	29,783	86	2·5
Devonshire Marble . . .	2·717	11,708	34,938	114	3·44
Bangor Slate . . .	2·859	15,510	41,590	144	4·51
Rowley Ragstone (basalt)	2·827	24,039	63,737	213	6·86
Aberdeen gray Granite .	2·678	16,868	51,123	155	4·44
Inverary Porphyry . . .	2·594	26,149	69,786	198	5·22

Thus with the ordinary sedimentary rocks the crushing weight (or that at which the blocks yield to pressure) is from 2¼ to 15½ tons per square inch of surface, while for the crystalline rocks it rises to 31 tons. The heat produced on the metal surroundings by the crushing was in most cases easily perceptible to the hand, and was so great in some of the granites and porphyries as to necessitate a delay for the apparatus to cool. Both Mallet and Rankin were of opinion that "in the crushing of a rigid material such as rock, almost the entire mechanical work (with the small residue of external work) reappears as heat." If, therefore, the disturbance affecting the

[1] *Phil. Trans.*, vol. 163 (1873), p. 147.
[2] Omitting fractions of a degree.

massive strata of a great mountain range were abrupt an intense degree of heat might be developed. There is reason, however, to suppose that such movements were slow during long periods of time, and it was only when the tension had reached a certain point that fracture and disruption, accompanied by a more rapid motion, took place.

What the force of the pressure may have been in these cases is manifested by the compression of the strata in the Alps, by the extraordinary folds and inversions of the rocks, and by the vertical cleavage (a resultant of pressure) which the whole mass of rocks has undergone. We may illustrate this point by the following generalised section across the central axis of the Alps along the line of the St. Gothard tunnel.

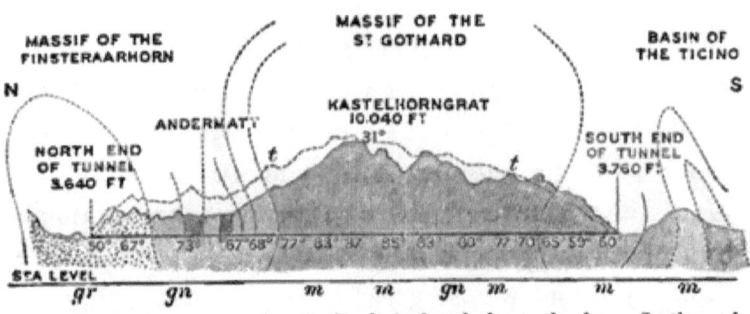

Fig. 5.—Section across St. Gothard (reduced from the large Section of Dr. Stapff).

gr. Granite ; *gn.* Gneiss ; *m.* Metamorphic schists ; *t.* represents the underground temperature curve.

But although the compression may have been excessive, and the actual mechanical displacement great, the crushing was not so complete nor so sudden as to produce the extreme effects suggested by Mallet. Complete crushing is not, however, necessary for our object, since the experiments show that on the first yielding of the rocks, which takes place when the weight is rather more than one third of the crushing weight, a large portion of heat is given off. Consequently as the heat would be developed gradually, and much of it might be dissipated during the long time that the disturbances may have lasted, the major effects obtained artificially would not be realised in nature. Nevertheless, although fusion may not have taken place, molecular and chemical changes were produced in the rocks indicative of the action of very considerable heat; and there is reason to believe that this heat was due to mechanical causes, rather than to the protrusion of the molten granitic centres. In fact M. A. Favre and other Swiss geologists now consider the granite in those ranges to have been in its present relative place when the elevation and crushing occurred.

Mallet further showed that the quantity of heat developed varied greatly in different rocks, and that, although compressed by the same force before their elastic limits were passed, yet, when released, it would render a quartz rock nearly three times as hot as a slate rock. Consequently granite and gneiss, with their large proportion of free quartz, would be more affected than most other rocks.

When, therefore, we consider at how late a geological period some of the great mountain chains have been uplifted, it is not impossible, looking at the magnitude of the *massifs*, that some residual portion of the heat produced by compression, faulting, and crushing may still exist in such modern chains as the

Alps, or in Continental areas of recent elevation, when that elevation has been accompanied by compression and faulting. This is a consideration which, although exceptional, should not be overlooked in the general question of underground temperatures, especially in mining districts, where we have to deal with disturbed areas, with their faults, dykes, and mineral veins; at the same time, there can be little doubt that the disturbances in these areas are generally of such high antiquity that there is, in most instances, small probability of the rock retaining much of the heat originally due to these causes.[1]

§ 7. Conductivity of the Rocks. Effects of Saturation and Imbibition

Although it is possible to account for many of the apparent discrepancies in the thermometric gradients by the causes discussed in the foregoing pages, yet it is evident that there are irregularities—not only between the three different groups of rocks, but also in each separate group respectively—which these causes do not adequately explain. As the rocks in each group are of very different petrological character, the differences of structure and composition must variably affect the conductivity of the rocks, and consequently the *thermic gradient.*

The researches of Professors Herschel and Lebour [2]

---

[1] For the heat due to this cause I have suggested the term "Regional Metamorphism." *Proc. Roy. Soc.*, vol. xl. (1885), p. 425.

[2] The results of their investigation are recorded in the reports of the British Association for 1874-1882.

in this country, and of M. Jannettaz[1] in France, have a special bearing on this subject. The former relate more especially to the differences dependent upon the petrological structure of the rocks, and the latter to those dependent on the component minerals. Tabulating the results of Professor Herschel and Lebour's experiments in accordance with the geological groups adopted in Tables II, III, and IV, we obtain the following mean conductivities of each class and group (table on p. 240).

Or dividing the list into the three groups under consideration, the results would be as under:

	Mean conductivity, k.	Mean resistance, r.
1. COAL MINES (*carboniferous strata*), Sandstones, Shales, Clays, &c.	0·00433	267
2. MINERAL MINES (*metamorphic and crystalline rocks*), Crystalline and Schistose rocks, Clay-slates, &c.	0·00473	225
3. ARTESIAN WELLS (*Secondary and Tertiary strata*), Sandstones, Oolites, Chalk, Greensands, Marls, &c.	0·00308	331

This shows a considerable difference between the conductivity of the rocks of the Mineral and Coal Mines with that of the strata in which the Artesian Wells are usually situated. But there are other conditions, hydro-geological and structural, which introduce many modifications affecting the value of these differences.

For example, in the Coal-measures, the coal-beds of which the conductivity is extremely small, may exercise more or less local influence, though these

---

[1] *Bull. Soc. Géol. de France*, 3rd ser., vol. iii., *et seq.*

Table of Thermal Conductivities, compiled from the Tables of Professors Z. S. Herschel and G. A. Lebour.

Nature of rock.		Absolute thermal conductivity, k.	Absolute thermal resistance, r.	Average. k.	Average. r.
Crystalline and Volcanic rocks.	Granite (mean of five varieties).	0·00584	172	0·00562	178
	Porphyry (Germany).	0·00513	195		
	Porphyritic trachyte.	0·00590	169		
	Basalt (Loch Katrine).	0·00560	179		
	Trap (Calton Hill).	0·00352	284	0·00475	221
	Serpentine.	0·00515	199		
Schistose rocks and Slates.	Gneiss (Germany).	0·00514	195		
	Mica schist (Scotland).	0·00520	192	0·00531	190
	Slates (three varieties)	0·00561	184		
	,, (across cleavage)	0·00395	253		
	Clay-slate (two varieties)	0·00327	307		
Sandstones.	Quartzite.	0·00954	105		
	Ganister sandstone.	0·00630	159		
	Craigleith.	0·00947	105	0·00734	139
	Hard sandstones.	0·00672	149		
	Micaceous flagstone (along cleavage).	0·00690	145		
	Micaceous flagstone (across cleavage).	0·00492	203		
	Soft red sandstone.	0·00397	252	0·00323	252
	New red sandstone	0·00250	...		
	,, ,, wet.	0·00600	166		
	Firestone (Upper Greensand).	0·00240	427	0·00172	689
	Quartzose sand.	0·00105	952		
	,, ,, wet.	0·00820	122		
Limestones, Oolites, &c.	Statuary marble.	0·00530	189		
	Devonian ,,	0·00645	157		
	Carboniferous limestone	0·00550	182	0·00561	180
	Magnesian ,,	0·00522	192		
	Oolite (Ancaster).	0·00370	270		
	Lias (building stone).	0·00360	278		
	Chalk.	0·00220	455		
Argillaceous strata.	Coal-measure shale.	0·00235	425	0·00242	411
	Clay (sun-dried).	0·00250	398		
	,, (the same wet and soft).	0·00350	270		
Mineral masses.	White quartz.	0·00957	104		
	Alabaster.	0·00360	278		
	Rock salt.	0·01280	78		
	Coal (Newcastle).	0·00068	1470		
	Cannel coal.	0·00120	787		

seams form but a small part of the entire mass. Thus there are—

	No. of workable seams.	Total thickness of coal.	Total thickness of coal-measures.
In the coalfield of Newcastle	16 ...	46 ft. ...	3,000 ft.
,, ,, N. Staffordshire	30 ...	130 ,, ...	5,000 ,,
,, ,, South Wales ...	75 ...	126 ,, ...	11,000 ,,

In the crystalline and metamorphic rocks, veins and layers of quartz and beds of quartzite are of common occurrence, though very irregular in their mode of distribution.

In the Artesian Wells gypsum and rock salt are of not unfrequent occurrence in the Triassic strata; in the remarkable instance of the Sperenberg boring, these minerals form a mass some hundreds of feet thick.[1]

The Influence of Water.—The above-named conditions are, however, subordinate to one of wider influence. The conductivity experiments of Messrs. Herschel and Lebour were, with few exceptions, made with blocks of dried rock. In a few instances they repeated the experiment with *wet* blocks of the same material, and with a remarkable difference in the result. The following indicates the extent of these differences:—

	Conductivity.	
	Dry.	Wet.
New Red Sandstone	0·00250	0·00600
Quartzose sand	0·00105	0·00820
Clay	0·00250	0·00350
Mean	0·00202	0·00590

Here we have substances which when dry present great thermal resistance, becoming when wet amongst

[1] It may, however, be a question whether it is not intercalated with some seams of gypsum.

the best of the rock conductors—equal, if not superior, to some of the crystalline and schistose rocks.

This condition becomes, in considering the question of conductivity in relation to underground temperatures, a matter of very considerable importance, for in nature underground dry rocks are the exception and wet rocks the rule, as the level of permanent saturation of the strata is regulated by the sea-level on one side, and by the level of the river valleys inland. In rocks above the line of permanent saturation the water of imbibition, or quarry water, is generally present, that being a property dependent on the capillarity of the rock; it is very strong in chalk and oolite, while it is slight in quartzose grits and limestones. There are therefore few rocks in which the influence of water is not felt.

The following is the proportion of water held in rocks when fully saturated :—[1]

	Complete saturation.
Granite (hornblendic)	0·06 in 100 parts.
,, fine grained	0·12 ,,
Basalt, Auvergne	0·33 ,,
Silurian slates, Angers	0·19 ,,
Devonian limestone	0·08 ,,
Coal shale	2·85 ,,
Coal-measures sandstone	14·30 ,,
New Red Sandstone	13·43 ,,
Inferior Oolite	23·98 ,,
Calcareous freestone, Paris	16·25 ,,
Chalk	24·10 ,,

In the hard granites, sandstones, and limestones, the water of imbibition differs but little in propor-

---

[1] These quantities are on the authority of the late M. Delesse *Bull. Soc. Géol. de France*, 2nd ser., vol. xix. p. 64.

tion from that of saturation. But in the sedimentary rocks the difference is more considerable, as the following examples will show:—

	Quarry water.
Gneiss, slightly decomposed	3·00
Plastic clay	19·56
Chalk	19·30

It is clear, then, that the conductivity of the underground rocks must, except possibly in some very hilly districts, be taken as that of the wet rock.

The conductivity even of coal will be increased, although the quantity of water that coal imbibes is very small. But unwrought coal also contains gas in a state of extreme condensation, or possibly in a state near liquidity, and this also may have an effect upon its conductivity.

Dip, Foliation, and Cleavage.—Messrs. Herschel and Lebour have also shown that the conductivity of slates varies accordingly as it is taken across or along the planes of cleavage—that while the conductivity along the planes of cleavage is equal to that of the crystalline rocks, it is no greater than that of soft sandstones across those planes. The lamination in micaceous sandstones produces a similar result:

	Conductivity.	Resistance.
Slates, along the planes of cleavage	0·00561	184
,, across ,, ,,	0·00395	253
Micaceous flagstone, along the laminæ	0·00690	145
,, ,, across ,, ,,	0·00492	203

M. Jannettaz[1] has extended the inquiry to a number of other rocks, and he shows that the

[1] *Bull. Soc. Géol. de France*, 3rd ser., vol. ii. p. 265; vol. iii. p. 499, *et seq.*

variation in conductivity in many rocks is largely dependent upon the presence of mica. He found that in a crystal of mica, heat was conducted about two and a half times more rapidly along the planes of cleavage than perpendicular to it. In augite these axes of the thermic curve are in the proportion of about two to one.

M. Jannettaz obtained results of a similar character, varying according to certain physical conditions, in a number of other minerals and in many rocks. The ratio of the two axes in the following rocks he found to be as follows:—

Gneiss of St. Gothard..........................	1 : 1·50
„ from near Chamouni	1 : 1·23
„ passing into mica schist............	1 : 1·63
Schists (triassic), St. Gervais	1 : 1·50
„ (carboniferous), Col Voza	1 : 1·80
Argillaceous schists	1 : 1·25
Cambrian Slate, Deville (Belgium)	1 : 1·86
Fissile micaceous limestone	1 : 1·31
Black and white limestone, Bonneville...	1 : 1·06

The thermic curves attain their maximum variation in talcose and micaceous schists and in slates. The greatest inequality, 1 : 3, was shown by a specimen of a talcose rock, of sp. gr. 2·7. The variation exists in all rocks showing schistosity or lamination, but in most ordinary stratified rocks the thermic curve remains that of the circle. It was found that the variation exists also in rock crystal, gypsum, felspar, &c. All the specimens experimented upon were dry.

It is obvious, then, that in gneissic rocks and slates, the *dip, cleavage,* and *foliation* may have a not unimportant effect on the conduction of heat. Lamination has a similar but lesser effect in argillaceous shales; in ordinary sandstones and limestones no such effects are produced. Whilst these effects therefore may be very manifest in the rocks generally associated with Mineral Veins, they can only be small in Coal Mines, although they may be in some places increased by a larger proportion of mica in the sandstones and shales.

Conclusions

The list of selected cases on which our conclusions are based may appear small, but the sources of error in experiments of underground temperatures are so many and so obscure, that without the fuller information which we have in these few instances, the larger number are not available for our purpose, though with the corrected data before named it may be possible to utilise some of them hereafter. To obtain the best approximation to the true thermometric gradient, I have therefore deemed it necessary to reject all the doubtful and more uncertain cases, so that in the case of the Coal Mines the list is limited to the eight instances given at p. 203; and in the Mineral Mines to eighteen of the seemingly most reliable rock and spring observations of Fox and Henwood (pp. 216, 218). The Artesian Wells give more uniform results, and of these I have selected fifteen, of which eight are overflowing wells, and seven not overflowing (pp. 228, 231).

Taking these three classes of observations, we obtain a mean gradient of 48·1:

	Thermometric gradient per 1° Fahr.
Artesian wells	51·5 feet.
Coal mines	49·1 ,,
Mines other than coal	43·2 ,,
Mean...	47·9

This is possibly the nearest approach that can be obtained to a true general normal gradient. In Coal Mines the effects of ventilation, and in other Mines the effects of chemical action and the circulation of water, leave yet many unsettled points; while in the case of Artesian Wells, the gradient of 51·5 feet may be too low in consequence of the unequal velocity of the water in deep overflowing wells, and of the uncertain measure of convection currents in those which do not overflow.

It seems at least evident that different geological areas have different thermometric gradients dependent on structural and hydrometric conditions. With regard to the latter, M. Delesse, as before mentioned, concluded that water might circulate to the depth of about eight miles before this limit was reached. This is probable, for the experiments of Regnault to determine the expansive force of the vapour of water up to a temperature of 239° C., prove that the pressure then is equal to $27\frac{1}{2}$ atmospheres; and though, beyond this, the estimate has only been carried by empirical formulæ, both experiment and calculation indicate that, with the increase of temperature, the increase of force is extremely rapid, and there is in all probability a point at which the vapour-tension of the heated water will equilibrate the hydrostatic pressure.

With respect to the possibility of change in the thermometric gradient at great depths, it is known that the conductivity of wrought iron diminishes as the temperature increases, and at a rate agreeing very closely with the empirical law that the conducting power of iron for heat is inversely as the absolute temperature. What the variation in rocks may be has yet to be determined experimentally;[1] we may presume it to exist, although it may differ materially in degree.

Therefore, taking into consideration the probable limitation of the percolation of water, and the possible diminution of conductivity with increase of depth, if there should be any alteration in the thermometric gradient, at great depths, it will be more likely to be in the direction influenced by these more or less certain factors; or in favour of a decreased conductivity and a more rapid thermometric gradient rather than otherwise.

I have made a few attempts to ascertain, with the data in our possession, whether there exists any indication of such variations within the limits of the depths reached, by comparing the gradients of the upper with those of the lower portions of the mines, but without arriving at any satisfactory result.[2] It is true that in the Coal Mines, taking a depth below

---

[1] The large proportion of iron present in the deeper seated igneous rocks is an element to be considered.

[2] The figures obtained are placed between the brackets in Tables II, III, and IV.

1,000 feet, the gradient, in all cases except two, shows, with increased depth, an increased rapidity, but it is a question whether this is not due to ventilation and convection currents causing too low a reading of the gradients in the upper part of the mines, and so throwing an apparent gain into the gradients in the deeper parts of the mines.

In the Mines other than Coal, some show at great depths a more rapid, and others a slower gradient; and it has to be observed that generally there is greater steadiness in the gradients at depths below 500 or 600 feet, than in those which are shallower.

In Artesian Wells and bore-holes, on the contrary, the gradient often appears to be more rapid in the upper than in the lower section of the wells, but this is clearly due to the action of convection currents; while the decrease in the diameter of the bore-hole with the increase of depth, by unequally checking the flow of water, differently affects the temperature of the water in the tubes as successive depths are reached. These latter considerations may however be omitted for the present.

P.S.—Some of my readers may be disposed to think that I have unnecessarily restricted the number of selected observations. As however the Tables include the complete list of 231 observations (exceeding, I believe, by more than one half, any previous list) it is open to any one to make his own selection and draw his own conclusions.

Table I.—GENERAL TABLE OF UNDERGROUND TEMPERATURES.

In this Table all the observations of which I could find record are placed in the order of date at which they were made. The discordance in those observations, which are repeated more than once, is generally due to the use of improved methods and instruments, or to corrections of the mean temperatures of the place. They are tabulated in the terms of the original papers, with the exception that the mean annual surface temperatures are corrected in accordance with later and better determinations. I am indebted to Mr. R. H. Scott, F.R.S., of the Meteorological Office, for some of these corrected temperatures (marked m), and for others to the lists of the Scottish Meteorological Society (marked s), and to the *Notices Scientifiques* of Arago (marked a). In other cases they are those given by the original observers or by Dove. The original surface heights have been corrected as far as possible. Where the temperature of the place has not been recorded, that of the nearest place of observation is given, allowance being made for difference in height, &c. The thermometric gradient and other particulars will be found in the classified Tables II, III, and IV.

By an Artesian Well is meant a bore-hole carried down to a deep-seated spring, which rises by that means to and over the surface. By Bore-hole is meant a boring of the like description, made either in search of minerals or of water, but without overflowing water.

A few ordinary surface wells are introduced merely as guides to the mean temperature at surface or at small depths.

I	II	III	IV	V	VI	VII
Locality.	Nature of Mine or Artesian Wells.	Height of Surface above Sea-level.	Mean Surface Temperatures.	Depth below Surface.	Temperature at Depth.	References and Remarks.
		Feet.	Fahr.	Feet.	Fahr.	
1. Giromagny, nr. Belfort.	Copper	1535	47°?	332	53·6°	Gensanne, 1740; Arago, *Notices Scientifiques*, vol. iii. p. 317 (1856). Temp. of Mulhouse, 51° F.
,, ,,	,,	...	,,	675	55·4	
,, ,,	,,	...	,,	1010	66·2	
,, ,,	,,	...	,,	1420	72	
2. Bex, Switzerland	Salt*	?	45?	721	63·5	Saussure, 1796; Arago, *op. cit.*
3. Guanaxuato, Mexico	Silver	6632	61?	1712	98†	
4. Cabrera ,,	,,	8510	60?	164	63	Humboldt, quoted by Arago, *op. cit.*, p. 338.
5. Tehuilotepec ,,	,,	5776	63?	358	75·5	
6. Micuipampa, Peru	,,	11614	46?	1500?	67	
7. Freiberg, Saxony— Beschertglück	,,	1378	43?	722	54·5	
,, ,,	,,	,,	,,	870	58	
,, ,,	,,	,,	,,	984	60	
8. ,, Himmelfahrt	,,	577	46?	328	50	Daubuisson, *Journal des Mines*, vol. xiii. p. 113 (1803); *Traité de Géognosie*, 1819, p. 444. Temp. of Dresden, 47° F.
,, ,,	,,	,,	,,	590	54·5	
,, ,,	,,	,,	,,	870	58	
9. ,, Kuhschacht	,,	656	45?	870	57	
10. ,, Junghöhe-Birke	,,	1050	44?	686	57·2	
,, ,,	,,	,,	,,	804	59	
,, ,,	,,	,,	,,	936	61	
,, ,,	,,	,,	,,	1082	62·5	
11. Brittany, Poullaouen	Lead & silver	348	51	246	53·5	
,, ,,	,, ,,	,,	,,	412	56	
,, ,,	,, ,,	,,	,,	489	58	Daubuisson, *op. cit.*, vol. xxi. p. 119 (1807).
12. ,, Huelgoet	,, ,,	568	50	230	54	
,, ,,	,, ,,	,,	,,	459	62·5	
,, ,,	,, ,,	,,	,,	781	66	

* The mine had been disused for three months.
† Temperature of spring issuing from lode. The air of the working galleries was 92° Fahr.

I	II	III	IV	V	VI	VII
Locality.	Nature of Mine or Artesian Wells.	Height of Surface above Sea-level.	Mean Surface Temperature.	Depth below Surface.	Temperature at Depth.	References and Remarks.
		Feet.	Fahr.	Feet.	Fahr.	
13. Freiberg, Saxony	Silver	1300 ?	43° ?	200	48·2°	De Trebra, 1805-7, *Ann. des Mines*, vol. i. p. 377 (1816), and vol. iii. p. 59. Obs. made in glazed niches in rock. The mean of 2 years' obs. No working going on.
,, ,,	,,	,,	,,	558	55	
,, ,,	,,	,,	,,	886	59	
,, ,,	,,	,,	,,	1246	66	
14. Whitehaven	Coal	50—100	‡ 48·5	480	60	R. Bald, *Phil. Jour.*, vol. i. p. 135 (1819). 48·5° is the temperature of Cockermouth.
15. Workington	,,	,,	,,	504	60	
16. Percy Main	,,	,,	,,	900	68	
17. Killingworth	,,	,,	,,	1200	74	
18. Cornwall	Copper & tin.	100—500	50 *	500—550	65	Dr. Forbes, "Temperature of Mines," *Trans. Roy. Soc. Cornwall*, vol. ii. p. 159 (1820). Average of observations made in six mines. Gives the temperature of the air and water. These are the water temperatures. [1] Here there was a strong current of air.
,,	,,	,,	,,	600—650	63	
,,	,,	,,	,,	700—750	65	
,,	,,	,,	,,	800—850	66	
,,	,,	,,	,,	900—950	71 [1]	
,,	,,	,,	,,	1150—1260	71	
,,	,,	,,	,,	1260—1350	74	
,,	,,	,,	,,	1350—1400	79	
19. ,, Dolcoath	,,	,,	280	50	240—300	58	R. W. Fox, *Trans. Roy. Soc. Cornwall*, vol. ii. p. 19 (1820). [2] Here were strong currents of air. Obs. in rock, except the last, which was in water.
,, ,,	,,	,,	,,	540—600	59 [2]	
,, ,,	,,	,,	,,	720—780	63	
,, ,,	,,	,,	,,	1140—1200	64 [2]	
,, ,,	,,	,,	,,	1320—1380	78 †	
,, ,,	,,	,,	,,	1380—1440	82	
20. ,, Huel Vor	Tin	...	51	6—60	52	Ibid. These are water temperatures. [3] Here the temperature of the air in gallery was 72°.
,, ,,	,,	,,	,,	180—240	61	
,, ,,	,,	,,	,,	480—540	63	
,, ,,	,,	,,	,,	600—660	64	
,, ,,	,,	,,	,,	600—720	66	
,, ,,	,,	,,	,,	720—780	70	
,, ,,	,,	,,	,,	780—840	69 [3]	
21. ,, Huel Damsel	Copper	...	,,	300—360	61	Ibid. These are air temperatures.
,, ,,	,,	,,	,,	480—540	69	
,, ,,	,,	,,	,,	600—660	70	
,, ,,	,,	,,	,,	720—780	70	
,, ,,	,,	,,	,,	780—840	73	
,, ,,	,,	,,	,,	840—900	70	
22. Neath, South Wales ‡	Coal	150 ?	‡ 50	540	62	

* In Cornwall a mean surface temperature of 50°, 51°, 52°, or even 53° F. was adopted in the early underground observations, and thermometric gradients were calculated on those several different scales. The more recent observations of mean annual temperature give for Penzance 51·5°, Truro 52°, Falmouth 51·4°, while Plymouth is 51·3°, and the high ground of Dartmoor 45·8°. We may take the mean annual temperature of the mining districts at 51°. Mr. R. Were Fox was, however, of opinion that the mean surface temperature of Cornwall was under 51°, and possibly even less than 50°. This will account for the apparent discrepancy between the gradients of many of the original observers and those given in these tables.

† A subsequent observation (No. 73) of Mr. Fox, made a year later, at the depth of 1,380 feet, gave a lower reading. A thermometer 4 feet long was placed in a hole 3 feet deep, at a spot where no workmen were employed, and where the current of air was small. The hole was filled with clay round the stem of the thermometer, which was left in that situation for eighteen months, and was found always to indicate a temperature of 76° or 76½°. In the experiment of 1820 the thermometer was buried in the rock to the depth of only 6 or 8 inches, and filled round with earth.

‡ The thermometer buried for some hours 1 to 2 feet under the ground.

I	II	III	IV	V	VI	VII
Locality.	Nature of Mine or Artesian Wells.	Height of Surface above Sea-level.	Mean Surface Temperatures.	Depth below Surface.	Temperature at Depth.	References and Remarks.
		Feet.	Fahr.	Feet.	Fahr.	
23. Carmeaux (Tarn)	Coal	820	55°?	20	55°	Cordier, *Essai sur la Température de l'Intérieur de la Terre*, 1822. Temp. of the air in galleries at Carmeaux 23·5° C., and at Littry 21° C. ¹ The mean surface temp. was estimated from shallow wells adjoining the coal pits.
,, ,,	,,	,,	,,	38	55·5	
,, ,,	,,	,,	,,	597	63	
,, ,,	,,	,,	,,	630	67	
24. Littry, Calvados	,,	107	51*	325 ¹	61	
25. Decise, Nièvre	,,	492	55?	29	54·5	
,, ,,	,,	,,	,,	351	64	
,, ,,	,,	,,	,,	561	72	
26. Cornwall, Huel Alfred	Copper	...	50	930	70	R. W. Fox, *Trans. Roy. Soc. Cornwall*, vol. iii. p. 313 (1828). ² Air 80°.
27. ,, Dolcoath	Copper & tin	280	50	1440	82 ²	
28. ,, Huel Trumpet	Tin	...	50	768	65	
29. Pregny, Geneva	Art. boring	308†	48‡	100 †	51·6	De la Rive and Marcet, *Mém. Soc. Phys. Geneva*, vol. vi. p. 503 (1833). Thermometer protected against pressure. This depth is equal to 713 English feet. Water not overflowing.
,, ,,	,,	,,	,,	209	53·2	
,, ,,	,,	,,	,,	460	57·4	
,, ,,	,,	,,	,,	600	61·5	
,, ,,	,,	,,	,,	650	62·7	
30. Sunderland	Coal	87	47·5	1584	72·6§	Phillips, *Phil. Mag.*, vol. vi. p. 446 (1834).
31. Tours	Art. well	180	53?	460	63·5	Arago, *Notices Scientifiques*, vol. iii. p. 347, *et seq.*
32. Lille (St. Venant)	,,	79	a 50·5	329	57·2	
33. Aire, Pas de Calais	,,	50?	49·7?	205	55·9	Arago, *Notices Scientifiques*, vol. iii. p. 347, *et seq.* ³ Temp. of Bochum, which is some miles further south, is 48·6°. The boring was ultimately carried to a depth of 2,113 feet.
34. Neu Salzwerk, Westphalia	Art. well salt	270	m 48 ³	787	70·7 ‖	
Neu Salzwerk	,,	,,	,,	1033	73	
,, ,,	,,	,,	,,	1673	81·5	
,, ,,	,,	,,	,,	2038	88·3	
35. Sheerness	Art. well	10	m 49·2	361	60	
36. Paris, École Militaire	,,	213	a 51	568	61·5	Walferdin, *Comptes rendus*, 1836, p. 314.¶
37. ,, Grenelle	,,	,,	a 51 **	568	...	Walferdin, *Comptes rendus*, 1837, p. 977; and Arago. Corrected to 1884.
,, ,,	,,	,,	,,	1312	74·7	
,, ,,	,,	,,	,,	1656	79·6	
,, ,,	,,	,,	,,	1797	83	
38. Cornwall, Levant	Copper & tin	80	50	1380	80	Henwood, *Report Brit. Assoc.*, 1837. Thermometer buried in rock
39. ,, Tresavean	Copper	362	50	1572	82	
40. ,, Consolidated	,,	318	50	1740	85·3	

* Temp. of Rouen is 50·7°.
† Above Lake of Geneva.
‡ Temp. of Geneva, 48·4°; height above sea, 1,335 feet.
§ Observation made in hole filled with water.
‖ The temperature of water at outflow in the first three observations did not agree with the temperatures at depth.
¶ All Walferdin's observations were made with overflow thermometers protected against pressure.
** Temp. of Paris. Another datum line of invariable temperature (53° F.) at the depth of 28 mètres (92 feet) in the cellars of the Paris Observatory, is sometimes taken.

I	II	III	IV	V	VI	VII
Locality.	Nature of Mine or Artesian Wells.	Height of Surface above Sea-level.	Mean Surface Temperatures.	Depth below Surface.	Temperature at Depth.	References and Remarks.
		Feet.	Fahr.	Feet.	Fahr.	
41. Rudersdorf, Berlin	Art. well	153	m 48·3°	880	74·3°	Bischof, *Edin. New Phil. Mag.*, vol. xxiv. p. 132 (1838).
42. St. Sever, Rouen	Art. boring	128	m 50·7	600	63·7°*	Girardin, *Comptes rendus*, 1838, p. 507.
43. St. André (Eure)	Art. well	...	51	246	54·5	Walferdin, *Comptes rendus*, 1838, p. 503.
,, ,,	,,	...	,,	830	64·4	
44. Yakoutsk, Siberia	Well	525	m 13	50	18·5	Erman, *Comptes rendus*, 1838, p. 501.
,, ,,	,,	,,	,,	77	19·6	
,, ,,	,,	,,	,,	119	23	
,, ,,	,,	,,	,,	182†	31	
45. Scotland, Carse of Falkirk	Art. well	...	s 46·5	231	51·5	R. Paterson, *Edin. Ph l. Mag.*, vol. xxvii. p. 71 (1839).
,, ,,	,,	...	,,	270	51·5	
,, ,,	,,	...	,,	380?	53	
,, Midlothian	,,	...	,,	...	...	Only the rate of increase given.
46. Cosseigne-les-Luxembourg	Art. boring	...	47	1105	78	Biver, *Comptes rendus*, vol. x. p. 41 (1840).
47. Cornwall	Tin & copper.	...	50	354	...	Fox, *Brit. Assoc. Rep.*, p. 310 (1840). Average of 53 mines. Only rate of increase given. In these cases see Table III.
,,	,, ,,	,,	,,	438	...	
,,	,, ,,	,,	,,	684	...	
48. Troyes, Aube	Art. well	359	a 52·3	410‡	60	Walferdin, *Bull. S.e. Géol. France*, vol. xi. p. 29 (1840).
48a. St. Ouen, Paris	,,	?	51	216	55·3	Arago, *Notices Scientifiques*.
49. Alfort, Marne	,,	...	51·5	177§	57·2	Lassaigne, *Comptes rendus*, Oct. 1842.
49a. Cornwall, East Wheal Crofty §	Copper	...	50	480	61	Small stream from lode.
				810	70·7	Small stream from rock.
49b. ,, Consolidated.	,,	318	,,	1704	89	Moderate stream from lode-end.
,, ,,	,,	,,	,,	1764	92·5	Large stream from lode-end.
49c. ,, United Mines	,,	,,	,,	1080	74	Very large stream from cross-course.
,, ,,	,,	,,	,,	1260	89·5	Moderate stream from lode-end.
49d. ,, Great Wheal Fortune	,,	,,	51	804	70	Large stream from rock.
				864	73	
49e. ,, Marazion	Copper & tin.	...	,,	222	56·5	Moderate stream from lode-end.
,, ,,	,,	,,	,,	480	63	Large ditto.
,, ,,	,, ,,	,,	,,	600	66	Small ditto.
49f. ,, Wheal Trenwith	Copper	...	,,	180	55·5	Small stream from lode.
,, ,,	,,	,,	,,	660	66	Large ditto.

* Thermometer not protected, but pressure allowed for; remained down 16 hours.
† Arago gives a depth of 377 feet with a temp. of 31° F. Ground frozen.
‡ The last 107 feet were obstructed, so that the actual depth was 517 feet.
§ Water overflows.

. 49a to 49p are from Mr. Henwood's paper, *Trans. Roy. Geol. Soc. Cornwall*, vol. v. pp. 389—402.

I	II	III	IV	V	VI	VII
Locality.	Nature of Mine or Artesian Wells.	Height of Surface above Sea-level.	Mean Surface Temperatures.	Depth below Surface.	Temperature at Depth.	References and Remarks.
		Feet.	Fahr.	Feet.	Fahr.	
49g. Cornwall, South Roskear, Camborne	Copper	...	50°	702	62°	Moderate stream from lode.
,, ,, ,,	,,	,,	,,	774	68	} Small ditto.
				834	71	
49h. ,, North Roskear.	,,	,,	,,	402	61	
,, ,, ,,	,,	,,	,,	642	66	
,, ,, ,,	,,	,,	,,	786	60·7	} Small ditto.
,, ,, ,,	,,	,,	,,	822	73	
49i. ,, East Pool	,,	,,	,,	372	59	Large ditto.
,, ,, ,,	,,	,,	,,	...	58·5	{ Large stream from cross-course.
49j. ,, Wheal Uny, Redruth	Tin	...	,,	399	58·3	Large stream from lode.
				432	60	Large stream from rock.
,, ,, ,,	,,	,,	,,	486	61·5	Small stream from lode.
49k. ,, Chacewater, Redruth	,,	,,	,,	768	75	Large ditto.
				,,	68	Small ditto.
,, ,, ,,	,,	,,	,,	,,	72	Small stream from rock.
49l. ,, East Wheal Virgin Consolidated Mines, Redruth.	Copper	,,	,,	1500	86·5	Large stream from lode end.
				1722	94·5	Large stream from rock.
				,,	92	Small stream from vein.
,, ,, ,,	,,	,,	,,	,,	91	Hole in rock.
49m. ,, Wheal Towan, S. Agnes	,,	,,	,,	648	62	Moderate stream from lode.
				804	70	Small stream from rock.
,, ,, ,,	,,	,,	,,	924	72	Small stream from lode.
49n. ,, Wheal Prudence	,,	,,	,,	654	65·5	Large ditto.
49p. ,, Wheal Vor	Copper & tin.	...	51	1420	80·5	} Temperature of rock.
,, ,, ,,	,,	,,	,,	1706	91	
50. ,, Binner Downs	Copper	...	50	500	56·5	
,, ,, ,,	,,	,,	,,	756	67	
,, ,, ,,	,,	,,	,,	816	65	
,, ,, ,,	,,	,,	,,	936	74·5	Henwood, Trans. Roy. Geol. Soc. Cornwall, vol. v. p. 389 and 402 (1843). The observations in these mines were taken in springs issuing from the lode or rock.
,, ,, ,,	,,	,,	,,	1056	82	
51. Devonshire, Wheal Friendship	,,	,,	,,	282	55	
,, ,, ,,	,,	,,	,,	450	54	
,, ,, ,,	,,	,,	,,	690	64	
,, ,, ,,	,,	,,	,,	810	69·5	
52. Cornwall, St. Ives Consols	Copper & tin.	...	51	108	57	
,, ,, ,,	,,	,,	,,	642	60·5	
,, ,, ,,	,,	,,	,,	462	65	
,, ,, ,,	,,	,,	,,	810	71	
53. ,, Wheal Wreath.	Tin	...	,,	162	53	} Ibid., p. 387.
,, ,, ,,	,,	,,	,,	1342	70	
,, ,, ,,	,,	,,	,,	1422	71·5	
,, ,, ,,	,,	,,	,,	,,	75	
,, ,, ,,	,,	,,	,,	1482	76	
54. Cornwall and Devon	Tin & copper.	...	50	180	54·3	
,, ,, ,,	,,	,,	,,	432	60·6	Ibid., p. 402. Average of various mines in ten districts.
,, ,, ,,	,,	,,	,,	762	67·4	
,, ,, ,,	,,	,,	,,	1050	78·6	
,, ,, ,,	,,	,,	,,	1440	85·5	
55. Monte Massi, Tuscany *	Shaft (?)	174	56?	1122	103	{ Matteucci, Comptes rendus, 1843, p. 937; 1845, p. 816.
,, ,, ,, ,,	,,	,,	,,	1214	107	

* No water. Shaft well ventilated. An abnormal centre of heat in this district.

I	II	III	IV	V	VI	VII
Locality.	Nature of Mine or Artesian Well.	Height of Surface above Sea-level.	Mean Surface Temperature.	Depths below Surface.	Temperature at Depth.	References and Remarks.
		Feet.	Fahr.	Feet.	Fahr.	
56. **Neuffen**, Wurtemberg *	Bore-hole......	1378	46°?	1263	101°	Daubrée, *Comptes rendus*, 1845.
57. **Mondorff**, Luxembourg †	Art. well	672	47?	1476	‡ 74·7	Rivot, *Ann. des Mines*, vol. viii. p. 79 (1845).
,,	,,	,,	,,	2200	93·2	
,,	,,	,,	,,	2297	? §	
58. **Mons** (Couchant de Flénu)§	Coal	...	...	...	...	*L'Institut*, April, 1845.
59. **Eastern Virginia**, U.S., Mills's Pit ‖	,,	...	56·7	420	63	
60. **Eastern Virginia**, U.S., Wills's Pit ‖	,,	...	,,	386	62	Professor H. D. Rogers, 1846. *Report on Coal Mines of Eastern Virginia.*
,,	,,	...	,,	570	65·5	
61. **Eastern Virginia**, U.S., Midlothian Pit	,,	...	,,	330	61·7	
,,	,,	...	,,	600	66·2	
,,	,,	...	,,	780	68·7	
62. **Meaux**, Marne	Art. well	...	51	230	57·2	D'Archiac, *Histoire des Progrès*, vol. i. p. 77 (1847).
63. **Ostend**, Belgium	,,	20	m 50	967	71·6	Dewalque, *Bull. Soc. Géol. de France*, vol. xx. p. 235 (1849).
64. **Vienna**	,,	637	m 50	616	60·8	*Bull. Soc. Géol. de France.*
65. **Mondorff**, Luxembourg.	,,	584	47·3?	1647	¹ 78·3	Walferdin, *Comptes rendus*, 1853, p. 250. ¹ Temp. of first overflowing spring. ² Temp. of the mud in bore-hole; this is unreliable on account of convection currents.
,, ,, ,,	,,	,,	,,	2362	² 81·7	
66. **Charleston**, U.S.A.	,,	20	m 66	50	68	Hume, *Edin. New. Phil. Journ.*, vol. lvii. p. 178 (1854). Nothing said about protection against pressure or convection currents. Observations taken at every 100 feet.
,, ,, ,,	,,	,,	,,	100	68	
,, ,, ,,	,,	,,	,,	400	72	
,, ,, ,,	,,	,,	,,	1000	84	
,, ,, ,,	,,	,,	,,	1106	88	
67. **Conselica**, Ferrara, Italy ¶	,,	27	m 53·9	164	59	Scarabelli, *Bull. Soc. Géol. de France*, vol. xiv. p. 102 (1856).
68. **Creuzot** (Torcy), Saône et Loire **	Bore-hole......	1017	48·5	1817	81	Walferdin, *Comptes rendus*, 1857, p. 971.
69. ,, (Mouillonge)	,,	1052	,,	2077	100	

* The abnormal temperature in this boring is attributed by M. Daubrée to the proximity of masses of basalt of post-miocene age; and by which basalt the adjacent rocks have been altered. The surface temperature of Tubingen is 8·7° C. The temperature from 30 mètres downwards marked 88·7° C.
† The water overflowed from a spring met with at a depth of 450 mètres. Bore-hole continued to further depth of 736 mètres. Thermometer not protected.
‡ Temperature of the water which overflowed at this level.
§ Temp. not given: only the rate of increase of 1° C. in 33·25 mètres.
‖ At Mills's and Wills's pits the temperature given is that of the water collected at bottom of shafts.
¶ Water rose 2 mètres above surface.
** Eighteen protected thermometers used. The boring at Torcy had been suspended for six months. The observations at Mouillonge were made after one to three days' rest. The two borings are 1,500 mètres apart.

I	II	III	IV	V	VI	VII
Locality.	Nature of Mine or Artesian Wells	Height of Surface above Sealevel.	Mean Surface Temperatures.	Depth below Surface.	Temperature at Depth.	References and Remarks.
		Feet.	Fahr.	Feet.	Fahr.	
70. 1837. Cornwall { Par Consols }	Tin ...	...	51°	768	74° ⎫	R. Were Fox, *Brit. Assoc. Reports for 1857*, p. 96.*
71. ,, ,, ,,	Copper	...	,,	1248	84 ⎬	[1] Under the sea—gallery quite dry.
72. ,, ,, Botallock	,,	40	51	1128	[1] 79 ⎭	
73. 822. ,, Dolcoath	Copper & tin	280	50	1380	76	
74. 1857. ,, ,, ,,	,, ,,	...	,,	1632	73	R. Were Fox, *Brit. Assoc. Reports for 1857*, p. 96; *Coal Commission Report*, vol. ii. p. 211.
75. 1853. ,, ,, ,,	...	...	,,	,,	79·5	
76. 1853. ,, Levant...	Copper & tin	80?	51	1530	[2] 74	[2] Near bottom of shaft.
76a. ,, ,, ,,	,, ,,	...	,,	,,	[3] 87	[3] Not far under the sea.
77. 1857. ,, ,, ,,	,, ,,	...	,,	,,	[4] 85	[4] Far under the sea—no working going on.
78. 1837. ,, Tresavean	Copper	362	50	1572	82·5	[5] A copious spring of water here gave 93·5°.
79. 1853. ,, ,, ,,	,,	,,	,,	2112	[5] 90·5	[6] A hot spring in another lode.
80. ,, United or Fowey Consols	,,	...	51	1728	93	
81. 18 7. ,, ,, ,,	,,	...	,,	1530	...	
82. ,, ,, ,, ,,	,,	...	,,	,,	[6] 116 ⎫	Henwood, *Edin. Phil. Mag.*, N.S., vol. vii, p. 147 (1858). The mine is 1,500 feet deep; well ventilated. No water: rainless district. Obs. in holes in rock 2 feet deep.
83. Chili	Colorado	3056	...	288	64·8 ⎬	
,,	,,	,,	...	762	67·5	
,,	,,	,,	...	900	67	
,,	,,	,,	...	1362	72 ⎭	
,,	,,	,,	...	,,	74·5	Wormley, *Amer. Journ. Science*, 2nd Ser., vol. XXX. p. 106 (1860).
84. Columbus, Ohio, U.S.A.	Art. well	834	m 53·3	90	53	
,, ,,	,,	...	,,	2575	88	
85. Dukinfield, 1848	Coal shaft	...	48·‡	17	51 ⎫	Fairbairn: *Brit. Assoc. Reports*, 1861, p. 53. Observations made in holes on side of shaft; thermometer left from half an hour to two hours. The holes all dry and mostly in shale or rock. No mention of temperature of air in the shaft, or of the depth of the holes.
,, ,,	,, ,,	...	,,	711	58	
,, ,,	,, ,,	...	,,	909	60	
,, ,,	,, ,,	...	,,	1119	64 ⎬	
,, ,,	,, ,,	...	,,	1338	67	
,, ,,	,, ,,	...	,,	1734	72	
,, ,,	,, ,,	...	,,	2151	75	
86. ,, 1858	A new shaft	...	,,	502	58	
,, ,,	,, ,,	...	,,	924	60	
,, ,,	,, ,,	...	,,	1000?	62?	
,, ,,	,, ,,	...	,,	1401	66·5 ⎭	
87. Rehme, Westphalia	Art. well	...	48·6 §	2280	88?	Fairbairn, *Brit. Assoc. Reports*, 1861.
88. Louisville, Kentucky?	,,	450	m 55·7	2086	83·5	Delesse, *Revue de Géologie*, vol. i. p. 9 (1862).
88a. Sahara Desert	,,	...	...	...	...	Only the rate of increase given. See Table IV.

* Most of these experiments were made at or near the ends of the deepest levels of the mines. Casella's thermometers were used in the later experiments. They were placed in holes 15 to 20 inches deep in the rock, which were carefully closed with clay, tow, or cotton. Thermometer left in for ½ to 1 hour.
† Walferdin's thermometers in strong iron case were used, but without protection against convection currents; left down 28 hours.
‡ The surface temperature of Manchester is M 48·6°. **Dunkinfield stands higher.**
§ This is the temperature of Boehum.
‖ Overflowing salt water; said to rise 52 mètres above surface.

I	II	III	IV	V	VI	VII
Locality.	Nature of Mine or Artesian Wells.	Height of Surface above Sea-level.	Mean Surface Temperatures.	Depth below Surface.	Temperature at Depth.	References and Remarks.
		Feet.	Fahr.	Feet.	Fahr.	
89. Naples, Largo Vittoria*	Art. well	...	m 59·9°	909	71·6°	Mallet, *Neapolitan Earthquakes*, vol. ii. p. 311 (1862).
90. ,, Royal Palace...	,,	...	,,	1460	68 †	
91. Ben Tallah, Algeria ...	,,	...	68 ? ‡	459	76	L. Ville, *Ann. des Mines*, 6th Ser., vol. v. p. 369 (1864). Overflowing wells.
92. Baraki ,, ,,	,,	...	,,	426	77	
93. Oued-el-Halleg ,, ,,	,,	...	,,	371	73·4	
94. Reggio, Italy	,,	...	...	65·5 §	2297		*Revue de Géologie*, vol. iii. (1864).
95. Ghadamés, Tripoli......	,,	...	73·4	394	84·2	Ditto ditto.
96. St. Petersburg	,,	15	m 38·6	525	50·5	*Ibid.* vol. iv. (1865), the water overflows.
97. Messis, Algeria	,,	...	68 ?	144	69·5	Degoussé et Laurent, *Revue de Géol.*, vol. iv. p. 25 (1865). Discharge of water per minute, respectively 150, 1,200, 15, 50, and 270 litres.
,, ,, ...	,,	...	,,	277	71·6	
98. Melahadalon, Algeria.	,,	...	,,	67	75	
,, ,, ,,	,,	...	,,	193	76	
,, ,, ,,	,,	...	,,	263	77·2	
99. Chega ,, ,,	,,	...	,,	138	73·4	*Ibid.*, p. 26 (1865).
100. Bothwell, Ontario	,,	...	45 ?	475 ¶	54	Sterry Hunt, *Chem. and Geol. Essays*, 1866, p. 159.
101. Rochefort, Charente Inférieure	,,	...	54·5	2812	111	Letter from Mauget and Lippmann, Paris, Jan. 1872. Overflowing mineral water.
102. Virac (Tarn)............	,,	795	55 ?	971	84·2	*Revue de Géol.*, vol. viii. (1869). Overflowing well.
103. Montigny, Belgium ...	Coal	...	51	2189	73·4**	W. Warington Smyth, *Quar. Journ. Geol. Soc.*, vol. xxiv. p. 81 (1·68).
104. Pendleton, Manchester	,,	126	m 48·6	1650	74	*Coal Commission Report*, vol. ii. pp. 90, 192, and 199. The work here had been open six years. Work here had been open six months.§§
,, ,, ,,	,,	,,	,,	1944	77	
,, ,, ,,	,,	,,	,,	2088	74††	
,, ,, ,,	,,	,,	,,	2214	86‡‡	
105. Hucknall Torkard Colliery, Notts............	,,	...	s 48·9‖‖	1250	70	*Royal Coal Commission Report*, vol. ii. p. 96. A new pit.

* The water rose above the surface.
† Attributes this low temperature to the influx of water at different depths.
‡ The mean temperature of Algiers is 64·5°.
§ Temperature of Messina.
‖ The rate of increase is given at 1° C. in 30 to 31 mètres. No other particulars.
¶ Water rose above surface, at rate of 700 gallons per hour, from the Corniferous Limestone.
** Temperature of air in gallery 68° F.
†† This was in the floor; temperature in coal 70° at a distance of 500 yards from the down-brow. In the same level 1,000 yards from the down-brow the temperature was 83° in the coal, and 82° in the floor.
‡‡ At 200 yards from the down-brow the temperature was 80° in the coal and 84° in the floor. At 400 yards, 82° in coal and 86° in floor. In a tunnel at the same level, the temperature of the shale was 76°; further on in fireclay it was 79°, and still further in hard rock 82°. Holes from 3 to 4 feet deep. Thermometers verified at Kew. Left 8 to 12 hours in holes perfectly dry.
§§ When open one year the reading gave 84°.
‖‖ This is the mean annual temperature of Nottingham.

ON UNDERGROUND TEMPERATURES—TABLE I

I	II	III	IV	V	VI	VII
Locality.	Nature of Mine or Artesian Wells	Height of Surface above Sea-level.	Mean Surface Temperatures.	Depth below Surface.	Temperature at Depth.	References and Remarks.
		Feet.	Fahr.	Feet.	Fahr.	
106. Annesley	Coal	...	* 48·9°*	1425	73°	*Royal Coal Commission Report*, vol. ii. p. 96. A new pit.
107. Kiveton Park	,,	...	,,	1200	71	Ditto ditto
108. Swanwick	,,	400	48	966	62·5	Ditto. Very wet shaft.
109. Moira, Warwickshire	,,	...	48·5	1030	...	Ditto ditto
110. Ruabon, North Wales	,,	420	48	1002	60 †	*Ibid.*, p. 104. Temp. of air in gallery ...... 58°
,, ,, ,,	,,	,,	,,	1503	70·5	Ditto ditto........ 58·5
,, ,, ,,	,,	,,	,,	1605	73	Ditto ditto......... 71
,, ,, ,,	,,	,,	,,	1770	78	Ditto ditto......... 71
111. Norley, Wigan	,,	157	,,	1049	75	Ditto ditto......... 71
,, ,,	,,	,,	,,	1184	75	Ditto ditto......... 72
,, ,,	,,	,,	,,	1282	78	Ditto ditto......... 73
,, ,,	,,	,,	,,	1487	80	Ditto ditto......... 79
						Distance from shaft.
						Ibid., p. 105, *et seq.* Temp. of air in gallery ......62° F. 27 yds.
112. Aberdare, Upper Duffryn,‡ No. 1 Station	,,	400	48·5	860	61	
,, ,, 2 ,, ·	,,	1220	...	1210	65	Ditto ditto....66 1587
,, ,, 3 ,, ·	,,	1330	...	1406	68	Ditto ditto....66 1877
,, ,, 4 ,, ·	,,	1540	...	1690	75	Ditto ditto....74 2327
113. ,, New Tredegar,§ No. 1 Station	,,	720	48 ?	865	58	Ditto ditto......60 110
,, ,, 2 ,, ·	,,	670	...	920	63	Ditto ditto......70 570
,, ,, 3 ,, ·	,,	1495	...	1673	69	Ditto ditto......71 2090
,, ,, 4 ,, ·	,,	1287	...	1549	67	Ditto ditto......72 1370
114. ,, Dowlais, Ironstone Mine	,,	1170	47	371	56	Ditto ditto.....55
,, ,, ·	,,	,,	,,	556	59	Ditto ditto......58
115. ,, Cwmbach	,,	480	48·5	250	55	
,, ,,	,,	1300	,,	985	61	
						Distance from shaft.
						Ibid., p. 123. Temp. of air in gallery 50° F. 312 yds.
116. Hetton, Durham	,,	400	47	1100	60 ‖	
,, ,,	,,	,,	,,	1155	66	Ditto ditto ... 69 1935
,, ,,	,,	,,	,,	1270	63	Ditto ditto ... 58½ 955
,, ,,	,,	,,	,,	1315	69·5	Ditto ditto ... 68 2980
,, ,,	,,	,,	,,	1360	66	Ditto ditto ... 64 1640
,, ,,	,,	,,	,,	1395	71	Ditto ditto ... 73 4332
,, ,,	,,	,,	,,	1400	70·5	Ditto ditto ... 72 3550
117. South Hetton, Durham	,,	204	...	1368	72¶	*Ibid.*, p. 123.
,, ,, ,, ·	,,	...	,,	1682	82	
,, ,, ,, ·	,,	...	,,	1938	96	

* This is the mean annual temperature of Nottingham.
† This observation was not taken until long after the pit was sunk, and coal evidently cooled.
‡ Beyond No. 2 station the circulation of air was partially stopped, and beyond No. 3 station entirely stopped, that part of the colliery having been abandoned for 18 months. The other Aberdare experiments were in collieries that were working at the time, and the air not shut off. The temperature taken 4 feet deep in the coal.
§ Although the surface at No. 3 station is 124 feet lower than at No. 4, it is further in the heart of the mountain, and the rise of surface at No. 4 is very abrupt. Temperatures taken 4 feet deep in coal, in dry holes and no gas.
‖ Holes in coal 3 feet deep, filled with water, and left 48 hours. **Thermometers then placed in them for 24 hours.**
¶ These temperatures were taken after boring operations had been suspended about a week; bore-holes full of water.

S

I	II	III	IV	V	VI	VII
Locality.	Nature of Mine or Artesian Wells.	Height of Surface above Sea-level.	Mean Surface Temperatures.	Depth below Surface.	Temperature at Depth.	References and Remarks.
		Feet.	Fahr.	Feet.	Fahr.	
118. Rosebridge, Wigan ...	Coal	157	48	483	64·5°	Coal Commission Report, pp. 143 and 188. All the holes were 1 yard deep and made air-tight. Holes allowed to stand 8 hours before thermometers were put in. Thermometer left 30 minutes.
,, ,, ,,	,, ...	,,	,,	564	66	
,, ,, ,,	,, ...	,,	,,	1674	78	
,, ,, ,,	,, ...	,,	,,	1815	80	
,, ,, ,,	,, ...	,,	,,	1890	83	
,, ,, ,,	,, ...	,,	,,	1989	85	Ibid., pp. 143 and 149. All the holes were 1 yard deep and made air-tight. Holes allowed to stand 8 hours before thermometers were put in. Thermometers left 30 minutes.
,, ,, ,,	,, ...	,,	,,	2037	87	
,, ,, ,,	,, ...	,,	,,	2202	88·5	
,, ,, ,,	,, ...	,,	,,	2235	89	
,, ,, ,,	,, ...	,,	,,	2283	91·5	
,, ,, ,,	,, ...	,,	,,	2322	91·5	
,, ,, ,,	,, ...	,,	,,	2349	92 ?	
,, ,, ,,	,, ...	,,	,,	2400	93 *	
119. Sharlston Colliery, Barnsley †	,, ...	240	,,	1005	65	Ibid., p 157. Distance from shaft. Air ....63° yds. 270
120. Victoria Colliery, Wakefield..................	,, ...	140	m 48·5	1455	78	Ditto ditto 65 600
121. Worthington Colliery, Lancashire	,, ...	...	48?	1808	82 ‡	Ibid., p. 194.
122. Ram's Mine, Lancashire, Pendleton Colliery §	,, ...	175	48·6	1044	72	Ibid., p. 199. At 300 yards from engine brow, 3 feet deep.
,, ,, ,,	,, ...	,,	,,	2088	78	At 500 yds, hole 3 ft. deep.
,, ,, ,,	,, ...	,,	,,	,,	82	At 1000 yds., hole 3 ft. deep.
,, ,, ,,	,, ...	,,	,,	2214	84	At 930 yds. from brow; hole 7 ft. 9 in. deep.
122a. Cornwall, Tresavean ‖	Copper & Tin	...	50	2130	99	Ibid., p. 85. Dry level hole, 4 ft. deep.
123. ¶ Blythswood, nr. Glasgow	Bore-hole......	?	m 47 **	60	47·9	Brit. Assoc. Report of 1869. Water in bore-hole.
,, ,, ,,	,,	,,	,,	180	50·5	
,, ,, ,,	,,	,,	,,	347	53·7	
124. South Balgray, Glasgow	,,	?	,,	60	48·2	Ibid. Original depth 1040 ft.; bore-hole silted up to 525 ft., and full of water.
,, ,, ,,	,,	,,	,,	180	51·1	
,, ,, ,,	,,	,,	,,	360	55·4	
,, ,, ,,	,,	,,	,,	525	59·5	
125. Carrickfergus, Belfast	Shafts	...	m 48·8 ††	570	62·4	Ibid A few feet of water in both shafts.
,, ,,	,,	,,	,,	770	66	

* The air at this depth **was** 18 to 22° lower than the rock temperature.
† The holes at this and **the** following pit were in coal and 2 yards deep, and perfectly dry. Thermometers left two days.
‡ This was the temperature of a brackish spring issuing from a girt rock 2' 9" thick met with in sinking the shaft, and discharging about 1600 gallons per hour.
§ These observations were taken two years after those of Mr. Knowles; the first three were made in holes 3 feet deep.
‖ *Coal Commission Report*, A 4, vol. ii.
¶ Nos. 123 to 165a are from the reports of Prof. Everett, *Trans. Brit. Assoc.*, 1869-83, inclusive.
** Mean annual temperature of Glasgow.
†† Mean annual temperature of Belfast.

ON UNDERGROUND TEMPERATURES—TABLE I

I	II	III	IV	V	VI	VII
Locality.	Nature of Mine or Artesian Wells.	Height of Surface above Sea-level.	Mean Surface Temperatures.	Depth below Surface.	Temperature at Depth.	References and Remarks.
		Feet.	Fahr.	Feet.	Fahr.	
126. Rosebridge, Wigan	Coal	157	48°	600	66°?	The temperatures (except the first) were taken during the sinking of the shaft by drilling a hole to the depth of a yard, plugging with clay, and leaving the thermometer ½ an hour.
,, ,, ,,	,,	,,	,,	1674	78	
,, ,, ,,	,,	,,	,,	1989	85	
,, ,, ,,	,,	,,	,,	2235	89	
,, ,, ,,	,,	,,	,,	2445	94	
127. St. Louis, U.S. America *		481	m 55	3029	107	Brit. Assoc. Report o 1870.
,, ,,		,,	,,	3843	105	
128. Mont Cenis Tunnel		9529	27·3†	5282	81·5	Brit. Assoc. Report of 1871.
129. Kentish Town, London	Art. well	187	49	395	56	These are the results ‡ of repeated observations commenced in 1869, by Mr. G. J. Symons, F.R.S. The rate of increase down to 910 ft. was 56 ft. for 1° F. Below that 49 ft. for each degree.
,, ,, ,,	,,	,,	,,	500	60	
,, ,, ,,	,,	,,	,,	700	62·8	
,, ,, ,,	,,	,,	,,	850	65	
,, ,, ,,	,,	,,	,,	1000	67·8	
,, ,, ,,	,,	,,	,,	1100	69·9	
130. Allenheads, Northumberland (Gin Hill Shaft)	Lead	1360?	s 44·2	340	49·3	Water stands in shaft at 328 ft. No reliance is placed on this determination.
,, ,,	,,	...	,,	390	51·2	
,, ,,	,,	...	,,	440	51·3	
131. ,, High Engine Shaft	,,	...	,,	857§	65·7	Water stands at 797 ft.
132. ,, Slit Mine	,,	...	45·3	660	65·1	Shaft full of water.
133. ,, Breckonhill Shaft	,,	1174	44?	42	46·5	Water stands 24 ft. down shaft. Unreliable.
,, ,,	,,	,,	,,	342	46·6	
134. Crawriggs, Glasgow	Bore-hole	200	47	50	47	
,, ,,	,,	,,	,,	200	50	
,, ,,	,,	,,	,,	350	51	
135. Moscow	,,	456	a 39·5	350	50	Records the same temperature at all depths.
,,	,,	,,	,,	994	50	
136. Durham, South Hetton	Coal	100?	47·5	1166 1	66	1 100 feet deep in bore-hole. Brit. Assoc. Report of 1872. Shaft is 1066 feet deep, and bore-hole 868 feet. Total depth 1924 feet.
,, ,,	,,	,,	,,	1466	72	
,, ,,	,,	,,	,,	1736	77½	
137. Paris, La Chapelle St. Denis	Art. well	...	51·5	328	59·5	Brit. Assoc. Report of 1873. The diameter of this bore-hole is 4 feet. Convection currents interfered with the results.
,, ,,	,,	...	,,	1312	69	
,, ,,	,,	...	,,	2165	76	
138. Stowmarket	,,	185	s 49·4	100	53	Original depth was 895 feet. Blocked. Uncertain results.
,,	,,	,,	,,	283	54	
139. Przibram, Bohemia ‖	Silver	?	44·7?	621	50·7	Brit. Assoc. Report of 1874. Observations made in holes 2 feet deep and far from the workings. Temperature of air not given.
,, ,,	,,	...	...	1209	58·3	
,, ,,	,,	...	...	1652	61·2	
,, ,,	,,	...	...	1900	61·4	

* Shaft to 71 feet, then a bore-hole.
† This is the estimated mean temperature of the surface summit level.
‡ Subsequent observations in 1879 established a temperature of 67·06° at 1,008 feet.
§ The thermometer could not be sunk below 857 feet, but the shaft extends to the depth of 957 feet.
‖ The section of this mine shows fifteen shafts. Herr Grimm attributes the slow increase of heat to the rocks, which are of Silurian age, being very quartzose.

I	II	III	IV	V	VI	VII
Locality.	Nature of Mine or Artesian Wells.	Height of surface above Sea-level.	Mean Surface Temperature.	Depth below Surface.	Temperature at Depth.	References and Remarks.
		Feet.	Fahr.	Feet.	Fahr.	
140. **Seraing**, Liége, Marie Colliery	Coal	177 ?	m 51° 4'	761	77°	Temperature of air in gallery 77½° F. Observations made in holes 5 metres deep. Thermometer left 24 hours.
,, ,, ,,	,,	,,	,,	1017	78	
141. ,, Henri Guillaume	,,	,,	,,	1656	87	
142. **Chiswick**, Middlesex	Art. well	25	49°·6	65	56·2	Brit. Assoc. Report of 1875. Water stands 60 feet from surface. 5 feet shaft down to 200 feet, then a bore-hole.
,, ,,	,,	,,	...	206	55	
,, ,,	,,	,,	,,	395	58	
143. **Swinderby**, Lincoln	Bore-hole	120 ?	48°·5 ?	100	63	Brit. Assoc. Reports of 1875–6. Springs at 790 and 950 feet. Strong convection currents affect results.
,, ,,	,,	,,	,,	500	68½	
,, ,,	,,	,,	,,	1000	69½	
,, ,,	,,	,,	,,	1300	70½	
,, ,,	,,	,,	,,	1500	73	
,, ,,	,,	,,	15°	1950	78	
,, ,,	,,	,,	,,	2000	79	
144. **Sperenberg**, Berlin	,,	?	m 48°·3 *	100 †	55	Brit. Assoc. Report of 1876. Diameter of bore at 3390 feet, 12 inches; then reduced to 6 inches. Bore-hole plugged to protect against convection currents, and observations corrected for pressure.
,, ,,	,,	,,	,,	700	70·8	
,, ,,	,,	,,	,,	1100	79·5	
,, ,,	,,	,,	,,	1400	84·5	
,, ,,	,,	,,	,,	1700	87·5	
,, ,,	,,	,,	,,	2100	96·3	
,, ,,	,,	,,	,,	3390	115·5	
,, ,,	,,	,,	,,	4052		
144a. **St. Petersburg**	Art. well	...	39 17	656	54	Water overflows.
145. **Anzin**, Nord, France, No. 1 Shaft	Coal	...	50°·5	126	56½	Obs. made in holes 0·6 to 0·7 mètres deep in sides of shaft during sinking. Little circulation of air; ½ hour elapsed between boring hole and inserting thermometer. Temp. of air in the wet shafts, 1, 2, and 3 was from 52° to 54°: in the dry shaft, No. 4, 59°.
,, ,,	,,	,,	,,	658	67½	
146. ,, No. 2 ,,	,,	,,	,,	286	55	
,, ,,	,,	,,	,,	607	63½	
147. ,, No. 3 ,,	,,	,,	,,	286	56	
,, ,,	,,	,,	,,	472	62½	
148. ,, No. 4 ,,	,,	,,	,,	99	70½	
,, ,,	,,	,,	,,	442	84	
149. **Schemnitz**, Hungary, Elizabeth Mine	Silver	from 1633 to 2504	from 47 to 42 (mean 44·3)	1368		Brit. Assoc. Report of 1877. The actual temperatures at depths are not given; only the rate of increase, which averages for the 5 mines 75½ ft. for each 1° Fahr.
,, Maximilian ,,	,,			430	...	
,, Amelia ,,	,,			985		
,, Stefan ,,	,,			715		
,, Siglisberg ,,	,,			1358		
150. **Boldon**, Newcastle	Coal	97	47°·1 §	1365	75	Still air in gallery 78½°. Trial holes 10 ft. deep; left four weeks to cool, temp. falling from 81 to 79°. Air travelled 3 miles and nearly stagnant.
,, ,,	,,	,,	,,	1514	79	
151. **Manegaon**, India	Bore-hole	1400	75 ?	10	81	The bore-hole had been 20 months at rest. Water stands to near top of tube.
,, ,,	,,	,,	,,	60	81	
,, ,,	,,	,,	,,	150	82·7	
,, ,,	,,	,,	,,	310	84·7	

* This is the mean temperature of Berlin.
† The depths here given are in Rhenish feet, 4,052 **Rhenish** = 4,172 **English** feet.
‡ In this shaft there was a seam of decomposing **coal at a depth of 90 mètres**.
§ Temperature of South Shields.

I	II	III	IV	V	VI	VII
Locality.	Nature of Mine or Artesian Wells.	Height of Surface above Sea-level.	Mean Surface Temperatures.	Depth below Surface.	Temperature at Depth.	References and Remarks.
		Feet.	Fahr.	Feet.	Fahr.	
152. Pontypridd, S. Wales	Coal*	554	49?°	855	62·7	*Report of* 1878. Hole in coal 4 ft. deep.
153. Bootle, Liverpool	Art. well	40?	49·6	226	52	Bore-hole at top 24 in. in diameter. Springs met with at depths of 318, 800, and 1,303 ft.
,, ,,	,,	,,	,,	1004	58·1	
,, ,,	,,	,,	,,	1302	59	
154. St. Gothard Tunnel	Tunnel †	...	31‡	3100	79·9	*Reports of* 1878-79. Swiss end of tunnel.
,, ,,		...	...	4101	83·8	
,, ,,	,,	...	...	4615	82·8	
,, ,,	,,	...	...	4965	83·7	Italian end of tunnel.
,, ,,	,,	...	...	4108	85·1	
155. Kingswood, Bristol, Deep Pit	Coal	216	m 50	441	54·7	Brit. Assoc. Report, 1879. Ventilation slight, and care taken to avoid air currents. Trial holes 2 ft. deep, plugged, and thermometer left twelve hours.
,, ,,	,,	,,	,,	1367	68·5	
,, ,,	,,	,,	,,	1769	74·7	
156. ,, Speedwell Pit	,,	216?	,,	1232	66·7	
,, ,,	,,	,,	,,	1439	69·7	
,, ,,	,,	,,	,,	1769	74·7	
158. Talargoch, Flintshire, 1st Pit	Lead	190	49·5	465	53·4	*Report of* 1880.—obs. distant from shaft 570 ft. Holes 2 feet deep
,, ,,	,,	,,	,,	555	52·9	Ditto ditto 321 ,,
,, ,,	,,	,,	49	636	58·8	Ditto ditto 2,522 ,,
,, ,,	,,	,,	,,	669	54	Ditto ditto 860 ,,
,, ,,	,,	,,	,,	1041	60·8	Ditto ditto 570 ,,
159. Dukinfield,§ Manchester	Coal	...	48	1987	74	Distance from air shaft, 1,380 ft.; air 71¼°.
,, ,,	,,	,,	,,	2407	80	Ditto, ditto 1,890,,; ,, 78½.
,, ,,	,,	,,	,,	2416	81	Ditto, ditto 1,800,, ; ,, 79.
,, ,,	,,	,,	,,	2700	86·5	Ditto, ditto 480,,; ,, 75¾.
160. Talargoch, Flintshire, 2nd Pit	Lead	190	40	600	62	*Report of* 1881. 1,200 ft. from shaft. Air still.
161. Ashton Moss, Manchester	Coal	...	48?	2790	85·3 ‖	Hole 3½ ft. deep.
162. Bredbury, Cheshire Nook Pit	,,	...	48·6	1020	62	The temperature of air in the galleries in these and preceding pit are not given.
	,,	,,	,,	1050	62·3	
163. Radstock, Somerset, Wells May Pit	,,	...	50	560	61·7	In holes 2 ft. deep. Distance from shaft and temp. of galleries not given. A moderate current of air was passing.
164. ,, Ludlow Pit	,,	...	,,	810	63	
	,,	,,	,,	1000	63	
165. Southampton Common	Art. well	140	,,	1210	69·7	Brit. Assoc. Report, 1883.
165a. Ballarat, Australia	Copper	...	...	760	72·5	Holes 3 ft. deep filled with water.

* The air current down the shaft between 20,000 to 30,000 cubic feet per minute. Neither the temperature of the air in gallery nor the distance from the shaft are given. In the other parts of the mine, the air currents showed differences of 2 to 3°, according to the season of the year.
 † The temperature of the springs in the tunnel was found to be higher than that of the rock.
 ‡ Temperature of surface on crest of tunnel.
 § These additional observations by Mr. Garside were made in the coal seams in holes 4 feet deep, and thermometer left 48 hours. The pit was entirely free from water.
 ‖ The gallery was free from any strong air current and the ground newly opened.

I	II	III	IV	V	VI	VII
Locality.	Nature of Mine or Artesian Wells.	Height of Surface above Sea-level.	Mean Surface Temperatures.	Depth below Surface.	Temperature at Depth.	References and Remarks.
		Feet.	Fahr.	Feet.	Fahr.	
166. Pitzbuhl, Magdeburg	Art. well ?	...	48'5 ?	495	...	De Lapparent's *Géologie*, 1881, p. 372.
167. Artern, Thuringia	,,	...	47'5 ?	1012	...	*Ibid.* The rate of increase only given.
168. Buda-Pesth, Hungary.	,,	...	a 50'5	190	52°	Communicated (1882) by Prof. Judd from letter of Prof. Szabó. The observations were made in 1877 and 1878. The high temperatures are dependent upon neighbouring old volcanic centres of activity. There are hot springs and trachytic rocks near.
,, ,, ...	,,	...	,,	216	64'4	
,, ,, ...	,,	...	,,	328	73'4	
,, ,, ...	,,	...	,,	1279	116	
,, ,, ...	,,	...	,,	1640	125	
,, ,, ...	,,	...	,,	1968	142	
,, ,, ...	,,	...	,,	2297	156	
,, ,, ...	,,	...	,,	2487	163	
,, ,, ...	,,	...	,,	2900	176	
,, ,, ...	,,	...	,,	2966	178	
,, ,, ...	,,	...	,,	3183	164	
169. Chicago	,,	600	m 45'9	751	55 *	*Smithsonian Inst. Report,* p. 240, 1874.
170. Chañarcillo, Chili	Silver	...	64	930	69'2	
171. Minas Giraës, Brazil	Iron	...	61'49	318	67'9	Henwood, *Trans. Roy. Soc. Cornwall,* vol. viii. p. 751 (1871).
172. Channel Islands, Sark and Herm	,,	...	a 51'6	222	55'5	
172a. Sheboyan, Wisconsin †	Art. well	620	...	1475	59'1	Chamberlain, *Geology of Wisconsin,* vol. ii. p. 165 (1873-77).
173. Ireland, various mines	Metalliferous	...	50 ?	342	53'4	
174. Cornwall, mean of ten mines	,,	103—300	...	672	68'8	
175. Wicklow, Ireland	,,	...	a 49'5	552	55'5	The observations were made in sumps or springs.
176. Waterford ,,	,,	...	50 ?	662	57'5	
177. County Cork ,,	,,	...	m 51 ?	840	61'5	
178. Villa Francisco Grande, Venice	Art. well	...	m 55'6	118	58	Laurent, *Revue de Géologie,* vol. xi. p. 258 (1875).
179. Venice, Gasworks	,,	...	,,	236	62'5	
180. Brussels, Belgium	,,	193	m 50'3	215	54	Vincent and Rutot, *Ann. Soc. Géol. Belg.*, vol. v. pp. 77 and 99 (1878).
181. Aerschot ,, ?	,,	...	,,	453	57'2	
182. Minden, Prussia	,,	238	m 48 §	2290	90'9	Raulin's *Géologie,* 1879, p. 84.
183. Arcachon, France	,,	...	55'4	413	61'9	
184. Pondicherry, East India	,,	...	a 82'5	261	93'7	Medlicott, *Geol. Survey of India,* 1881.
185. Dundee	,,	160 ?	m 46'6	238	50	*Sixth Report of Rivers' Pollution Commission,* 1868. Temperature of Wells.
186. Bradford, Yorkshire	Ord. well	366	a 47'8	360	54'5	
187. Blackburn ‖	Coal	347	48 ?	210	49'5	
188. Birmingham, well	Private well	340 ?	m 48'7	300	50	

* An infiltration of water from a higher level was suspected.
† Flowing water was obtained at 1,340 feet in the upper portion of **a Silurian sandstone**. Water rose 104 feet above the surface. Discharge of water = 225 gallons per minute.
‡ The water rose 5 metres above surface.
§ See temperature of Bochum, p. 251.
‖ Abandoned coal shaft.

I	II	III	IV	V	VI	VII
Locality.	Nature of Mine or Artesian Wells.	Height of Surface above Sea-level.	Mean Surface Temperatures.	Depth below Surface.	Temperature at Depth.	References and Remarks.
		Feet.	Fahr.	Feet.	Fahr.	
188a. Birmingham Waterworks	Ord. well	350	m 48·7	400	53·60	
189. Kidderminster	,,	320	49 ?	160	54	
190. St. Helens Waterworks	,,	100 ?	49	270	50	
191. Tranmere, Cheshire	,,	...	49·4*	428	50	
192. Wallasey ,,	,,	...	,,	246	51·8	
193. Worksop	,,	127	m 48·7	214	51·8	
194. Scarborough	,,	176	m 47·8	214	54	
195. Eastbourne	Art. well	25	m 50·0	160	50	
196. Deal Waterworks	Ord. well	20	50	115	52	*Sixth Report of Rivers' Pollution Commission,* 1868. Temperature of Wells.
197. Dover Castle	,,	380	m 50·3	367	55·4	
198. Dover Waterworks	,,	40 ?	,,	220	52	
199. Grimsby Docks	,,	10	48	300	52·5	
200. Deptford Waterworks	,,	31	a 50·3	250	54	
201. Sittingbourne	,,	50	49·5 ?	400	53	
202. Braintree	Art. well	220	49·5 ?	430	54	
203. Wimbledon	,,	170	49·6	200	54·3	
204. Carisbrooke Castle	Ord. well	...	50 ?	240	52·5	
205. Colchester	Art. well	109	m 49·4	400	52·6	
206. Trowbridge Waterworks	Ord. well	140	49·5 ?	200	52	
207. Ostend†	Art. well	10 ?	m 50	567	59	Letter from Prof. Dewalque; February, 1883.
,,	,,	...	,,	617	...	
,,	,,	...	,,	981	71·6	
208. Bourbourg,‡ Dunkirk	,,	...	50·2	544	59	Letters from Prof. Gosselet, February, 1883.
209. Dunkirk	,,	...	m 50·2	426	52	
210. Newport, I. of W.	,,	60	50·0 ¹	467	62	Letter from Mr. H. Turner, September, 1883. ¹ The temperature of Osborne.
211. Gosport	,,	20	m 50·6	372	55	
212. Bothwell,§ Ontario	,,	...	45 ?	475	54	Sterry Hunt, *Chem. and Geol. Essays*, p. 159.
213. Croix, Dept. du Nord ‖	,,	...	50 ?	271	75	Letter from M. Ortlieb, Feb. 1883. Average of 7 wells.
214. Mons, Grameries	Coal	344	50·5	679	66 ¶	Letter from M. F. L. Cornet, April, 1883. A great quantity of water flowed from the rocks in this pit.
,, ,,	,,	,,	,,	1013	69	
,, ,,	,,	,,	,,	1141	60	
,, ,,	,,	,,	,,	1825	58·5	

* 49·4 is the temperature of Chester.
† The water issued at the surface with a temperature of 18° C., and rose 8 mètres above sea-level at the first depth, and 11 mètres at the last depth.
‡ The well was carried through Tertiary strata to a depth of 250 mètres into the chalk.
§ The water rose above surface at the rate of 700 gallons per hour, from Corniferous limestone.
‖ These wells pass through Tertiary strata and end in Carboniferous limestone. The water rose them above the surface; one well delivers 12,000 litres per hour.
¶ Thermometer placed for not less than an hour in holes, 1 mètre deep, excavated in side of gallery.

I	II	III	IV	V	VI	VII
Locality.	Nature of Mine or Artesian Wells.	Height of Surface above Sea-level.	Mean Surface Temperature.	Depth below Surface.	Temperature at Depth.	References and Remarks.
		Feet.	Fahr.	Feet.	Fahr.	
215. Mons, La Louvière Pit.	Coal	410	50·5°	1194	76°	Letter from M. Cornet, April, 1883. Observations made 1200 mètres from the shaft in galleries; perfectly dry and not ventilated.
,, ,, ,,	,,	,,	,,	1456	81	
,, ,, another.	,,	251	,,	787	69 *	Ibid. In dry gallery 100 mètres from shaft.
,, ,, ,,	,,	,,	,,	1361	81	Ibid. In a new shaft without water.
216 Mons, Cuesmes Pit	,,	213	,,	1548	79·5	Ibid. At 2180 mètres from shaft.
,, ,, ,,	,,	,,	,,	,,	79·7	Ibid. At 2400 do. do.
,, ,, ,,	,,	,,	,,	,,	78·5	Ibid. At 2600 do. do.
,, ,, ,,	,,	,,	,,	1680	82	Ibid. At 700 do. do.
,, ,, ,,	,,	,,	,,	,,	81·5	Ibid. At 900 do. do. In galleries dry and not ventilated.
,, ,, ,,	,,	,,	,,	1709	85	Ther. placed in a spring of salt water issuing from a bed of sandstone.
217. North Seaton, N'castle†	,,	−50	47·5	660	61	Brit. Assoc. Report, 1883.
218. Ashton Moss, Manchester	,,	...	48	2889	84	Ther. left 48 hours in hole.
229. Dolcoath, Cornwall	Tin	280	50	252	64	
,, ,, ,,	,,	,,	,,	320	65	
,, ,, ,,	,,	,,	,,	876	67·8	Brit. Assoc. Report. 1883.
,, ,, ,,	,,	,,	,,	1118	65 ¹	¹ Observations considered defective.
,, ,, ,,	,,	,,	,,	1884	70 ¹	
,, ,, ,,	,,	,,	,,	2124	83	
,, ,, ,,	,,	,,	,,	2244	90	
221. Passy,‡ Paris	Art. well	158	51	1924	82·5	Letter from MM. Mauget and Lippmann, Jan. 1872.
223. La Fayette, Indiana	,,	...	...	213	55	Rev. Géol., vol. i. p. 9.
224. Buenos Ayres	,,	...	62·7	255	69·8	Quart. Journ. Geol. Soc., vol. xix. p. 69.
226. Croft, Whitehaven §	Coal	72	48·5	1140	73	
227. ,, ,,	,,	...	,,	1250		Brit. Assoc. Report for 1882.
228. Lye Cross, Dudley ǁ	,,	822	47·5	700	57·5	
229. Denton,¶ Manchester	,,	...	48	1370	66	
230. Richmond,** Surrey	Art. well	...	49·6	1176	70	Professor Judd in Quart. Journ. Geol. Soc. vol. xl. p. 724 (1884).
231. ,, ,,	,,	,,	,,	1337	75·5	

* In another trial made 300 mètres from the shaft, the temperature at the same depth was found to be 70·75° F.

† These observations were made by Professor Lebour at a point under the sea half a mile beyond low-water mark, and 620 feet below sea-bottom.   ‡ Diameter of well at top, 4 feet.

§ Hole 4 feet deep, bored upwards in roof of coal. The stations in this pit were under the sea.

ǁ Hole 4 feet deep in shaly floor, under the "10 yards" coal.

¶ This is in a valley near the Dukinfield pit.   ** Slight overflow, 4 or 5 gallons per minute.

Those observations in Table I. in which there are readily-apparent errors, or which are repeated more correctly at later dates, are not brought forward in the other Tables. Nor has it been considered necessary to repeat the fuller particulars there recorded in No. I., nor other observations, except such as bear upon the rate of increase of temperature with depth.

For subjects peculiar to the separate groups, special columns are introduced in Tables II, III, and IV.

TABLE II.—COAL PITS AND SHAFTS. (For other particulars see Table I.)

Those observations in **Table I.** in which there are obvious errors, or are repeated at a later date, are not brought forward.

In this table there is no separate column for the strata, as they all consist of the usual *shales, sandstones,* and *coals* of the *Coal-measures*. In the few instances where they are overlaid by newer strata, the particulars are given in the notes.

Column IV. gives the distance in yards of the place or station of observation from the shaft, showing the distance the air has to travel before reaching the face of the coal. It is, however, not often recorded.

Column V. gives the depth of the hole drilled in which the thermometer was placed, and whether in the *coal* (C), *rock* (R), or *water* (W).

In Column VI. the numbers in brackets show the difference of temperature between the two depths given in *italics* and bracketed in Column III., while the rate of increase *between these depths* is given also in *italics* and between brackets. Column VII. gives the temperature of the air in the gallery in which the rock or spring temperatures are taken. In Column VIII. the figures in thicker type refer to the gradient of the entire depth of the mine.

I	II	III	IV	V	VI		VII	VIII
Number in Table I.	Name of Colliery and Place.	Depth in Shaft.	Distance of Station from Shaft.	Depth and Position of Hole. Coal C. Rock R. Water W.	Temperature at Depth			Thermometric Gradient.
					in Trial Hole or Water.		of the Air in Gallery.	
					Fahr.		Fahr.	
	ENGLAND.	Feet.	Yards.	Feet.				Feet.
14	Whitehaven Colliery *	480	...	... W	60° W		...	42
15	Workington Colliery	504	...	... ,,	60 ,,		...	44
16	Northumberland, Percy Main ..	900	...	... ,,	68 ,,		70°	46
17	,, Killingworth.	1200	...	... ,,	74 ,,		77	47
150	Newcastle, Boldon Pit	1365	...	10 R	75 R		75.5	49
	,, ,,	1514	...	,, ,,	79 ,,		78.5	47
	,, ,,	(1365—1514)	...	... ,,	(4.0)		...	(37)
217	,, North Seaton	620	...	... ,,	61 ,,		...	45
30	Sunderland, Monkwearmouth ..	1584	...	... R	72.6 ,,		...	62
116	Durham, Hetton Pit	1100	312	3 C	60 C		50	85
	,, ,,	1135	1025	,, ,,	68 ,,		69	55
	,, ,,	1270	955	,, ,,	63 ,,		58.5	79
	,, ,,	1315	2030	,, ,,	69.5 ,,		68	59
	,, ,,	1360	1640	,, ,,	66 ,,		62	72
	,, ,,	1395	4330	,, ,,	71 ,,		73	58
	,, ,,	1400	3550	,, ,,	70.5 ,,		72	60
136	,, South Hetton †	1060	...	1000 R	66 R		...	60
	,, ,,	,,	...	400 ,,	72 ,,		...	59
	,, ,,	,,	...	670 ,,	77 ,,		...	58
	,, ,,	(1160—1736)	...	... ,,	(11.1)		...	(52)
	,, ,,	(1460—1736)	...	... ,,	(7.1)		...	(52)

* At Whitehaven and **Workington** the Coal-measures are unconformably overlaid by 200 to 300 feet of Red Sandstones and Marls.

† The temperature observations in this pit were made in a bore-hole drilled at the bottom of the shaft. The first series of observations were made (*Coal Commission Report*, vol. ii. pp. 128 and 133) after the boring operations had ceased twenty minutes. The temperature at the bottom of the bore-hole, then 858 feet deep, or 1,924 below the surface, was 96°. The experiments were repeated after the boring operations had been suspended about a week and the temperature found to be the same as before. But those made three years later and recorded in the *British Association Reports*, show a considerable decrease of temperature. The abandoned bore-hole had then silted up to the depth of 644 feet. The encased thermometer was pushed down to 26 feet in this, or to a depth of 1,736 feet from surface, where the temperature was found to be 77.1°.

I	II	III	IV	V	VI	VII	VIII
Number in Table I.	Name of Colliery and Place.	Depth in Shaft.	Distance of Station from Shaft.	Depth and Position of Hole. Coal C. Rock R. Water W.	Temperature at Depth — in Trial Hole or Water.	Temperature at Depth — of the Air in Gallery.	Thermometric Gradient.
	ENGLAND—*continued*.	Feet.	Yards.	Feet.	Fahr.	Fahr.	Feet.
121	Lancashire, Worthington.........	1803	...	... W	82° W	62°?	53
111	Wigan, Norley Coal Co............	1487	...	... R	80 R	70	46
		609	...	... ...	66	...	...
126	,, Rosebridge *............	1674	...	3 R	78 R	...	56
	,, ,, ...	2445	...	,, ,,	94 ,,	...	53
	,, ,, ...	(1074—2037)	...	... ...	(9'0)	...	(40)
	,, ,, ...	(2037—2445)	...	... ...	·7'0)	...	(55)
	,, ,, ...	(1074—2445)	...	... ...	(10'0)	...	(55)
104	Manchester, Pendleton †.........	1944	...	3 to 4 R ‡	77 ,,	64 . R ‡	68
	,, ,, ...	2214	400	,, ,,	86 ,,	67 ,,	59
	,, ,, ...	(1944—2214)	...	... ...	(9'0)	...	(30)
122	,, ,, ...	2088	500	3 R	78 ,,	65 R	70
	,, ,, ...	,,	1000	,, ,,	82 ,,	71? ,,	63
159	Dukinfield, Astley Pit §	1987	450	4 ,,	74 ‖	71'5 ,,	74
	,, ,, ...	2416	600	,, ,,	81 ,,	70'0 ,,	73
1·9	,, ,, (2nd obs.)	2700 ¶	160	,, ,,	86·5 R	75'5 ,,	70
	,, ,, ...	(1987—2700)	...	... ...	(12'5)	...	(51)
161	Manchester, Ashton Moss (1881)	2790	...	3½ R	85·3 R	...	74
218	Ashton Moss (1883).................	2880	...	,, ,,	84 ,,	...	82
162	Cheshire, Bredbury	1020	...	,, ,,	62 ,,	...	78
102	,, Nook	1056	...	,, ,,	62·3 ,,	...	78
119	Barnsley, Sharlston Pit	1005	270	6 ,,	65 ,,	63	63
120	Wakefield, Victoria Pit............	1455	600	,, C	78 C	75	50
105	Nottinghamshire, Hucknall Torkard **...	1250	...	2? ,,	70 ,,	68·5	60
106	,, Annesley **..	1425	...	,, ,,	73†† ,,	67	59
107	,, Kniveton Park **	1200	250	,, ,,	71 ,,	72'5 ?	55
108	,, Swanwick ‡‡.	966	...	,, ,,	62·5 ‡‡	...	72
109	Warwickshire, Moira §§	1030	...	,, ,,	66 ,,	,,	61
155	Bristol, Kingswood Pit	441	...	2 R	54·7 R	...	83
	,, Deep Pit.....................	1367	...	,, ,,	68·5 ,,	...	77
	,, ,,	1767	...	,, ,,	74·7 ,,	...	71
	,, ,,	(1367—1767)	...	... ...	(6·2)	...	(64)
229	Denton, Manchester	1317	...	4 ...	66	...	77
156	Bristol, Speedwell ‖‖	1232	...	1 R, C	66·7 C	...	78
	,, ,,	1439	...	2 ,,	69·7 ,,	...	74
	,, ,,	(1232—1439)	...	... ...	(3'0)	...	(69)

* The observations were made in **holes at bottom** of shaft during sinking.
† 340 feet of Triassic and Permian **strata overlie** the Coal-measures. The distances in Column IV. are from the down-brow.
‡ The temperature of the rocks in this pit was from 2° to 4° higher than that of the coals.
§ 708 feet of New Red Sandstone overlie the Coal-measures.
‖ The first series of observations at Dukinfield, although taken with care, give neither the temperature of the air in the shaft nor the depth and position of the holes. Mr. Dickinson also calls attention to the fact that before the shaft was sunk two of the principal seams of coal in the upper part had been worked away from the out-crop down towards the Astley shaft, and in one case a tunnel had been driven to where the shaft had come.
¶ This pit **has now (1884) been carried** to the great depth of 3,150 feet.
** These were **all new pits**.   †† Hole at bottom of shaft.   ‡‡ Very wet pit.
§§ An old pit.   ‖‖ The strata dip about 1 in 6.

ON UNDERGROUND TEMPERATURES—TABLE II

I	II	III	IV	V	VI	VII	VIII
Number in Table I.	Name of Colliery and Place.	Depth in Shaft.	Distance of Station from Shaft.	Depth and Position of Hole. Coal C. Rock R. Water W.	Temperature at Depth in Trial Hole or Water.	of the Air in Gallery.	Thermometric Gradient.
	ENGLAND—continued.	Feet.	Yards.	Feet.	Fahr.	Fahr.	Feet.
163	Bath, Radstock, Wells May Pit*	560	...	2 R	61·7 R	...	48
164	,, ,, Ludlow Pit	810	...	,, C	63 C	...	62
	,, ,, ,, ,,	1000	...	,, ,,	,, ,,	...	67
110	North Wales, Ruabon	1092	...	3 ,,	60 ,,	58°	91
	,, ,,	1563	...	,, ,,	70·5 ,,	56·5	70
	,, ,,	1605	...	,, ,,	73 ,,	71	68
	,, ,,	1770	...	,, ,,	78 ,,	,,	61
	,, ,,	(1092—1770)	...	,, ,,	(13)	...	(59)
112	S. Wales, Aberdare, Upper Duffryn †.	360 ‡	27	4 C, R	61 C, R	62	30
	,, ,,	1210	1587	,, C	65 C	65	58
	,, ,,	1400	1877	,, C, R	68 C, R	66	60
	,, ,,	1690	2327	,, ,,	75 ,,	74	54
	,, ,,	(1210—1690)	...	... ...	(8)	...	(60)
113	,, ,, New Tredegar †	855 ‡	110	4 C	58 C	60	80
	,, ,, ,, ,,	920	570	,, ,,	63 ,,	70	61
	,, ,, ,, ,,	1673	1370	,, ,,	67 ,,	72	80
	,, ,, ,, ,,	1549	2000	,, ,,	69 ,,	71	74
	,, ,, Vochriw Dowlais §	1103	66	,, ,,	60 ,,	59	80
	,, ,, ,, ,,	1320	466	,, ,,	62 ,,	,,	83
	,, ,, ,, ,,	628	2000	,, ,,	61 ,,	65	42
114	,, ,, Dowlais (ironstone)	371	160	,, R	56 R	55	61
	,, ,, ,, ,,	536	485	,, ,,	59 ,,	58	45
115	,, ,, Cwmbach	230	80	,, C	55 C	48	36
	,, ,, ,, ,,	988	1600	,, ,,	61 ,,	63	78
152	,, Pontypridd	855	...	,, ,,	62·7 ,,	...	62
22	,, Neath	540	...	2 R	42 R	...	45
226	Croft, Whitehaven	1140	430 ‖	4 ,,	73 ,,	...	47
227	,, ,,	1250	1340	,, ,,	,, ,,	...	51
228	Lye Cross, Dudley	700	...	,, ,,	57·5 ,,	...	70
	BELGIUM.						
140	Liège, Seraing Collieries¶	761	...	16 R	77 R	78	30
	,, ,,	1017	...	,, ,,	78 ,,	77	40
141	,, ,,	1656	...	,, ,,	87 ,,	77·5	46
	,, ,,	(761—1656)	...	... ...	(10)	...	(90)
	,, ,,	(1017—1656)	...	... ...	(9)	...	(70)

* Dip small. The Coal-measures in this district are covered unconformably by 100 to 200 feet of Jurassic and Triassic strata.

† The dip is small in these Aberdare pits.

‡ The depths in these pits are not the *depths of the shaft*, but are, in each pit, taken on one and the same level, and the depths given are those *beneath the surface*, the differences of depth being caused by the coal seam passing from the valley in which the shaft is situated under an adjacent hill.

§ In this pit the working proceeds from a hill towards a valley.

‖ These distances are those of the stations under the sea, which is there about 12 fathoms deep. Taking the temperature of the sea at 48°, and deducting 72 feet, Professor Everett makes the thermometric gradients 45° and 47°.

¶ The dip of the coal in the Liège pits is very rapid. In the *Brit. Assoc. Reports*, the temperature of the ground at a depth of 5 mètres is estimated at 54° F. The gradient here given is, however, calculated on a mean surface temp. of 51°.

I	II	III	IV	V	VI	VII	VIII
Number in Table I.	Name of Colliery and Place.	Depth in Shaft.	Distance of Station from Shaft.	Depth and Position of Hole. Coal C. Rock R. Water W.	Temperature at Depth — in Trial Hole or Water.	Temperature at Depth — of the Air in Gallery.	Thermometric Gradient.
	BELGIUM—*continued*.	Feet.	Yards.	Feet.	Fahr.	Fahr.	Feet.
214	**Mons**, Grameries Colliery *	679	...	3¼ R	66° R	...	45
	,, ,, ,,	1013	...	,, ,,	69 ,,	...	56
215	,, La Louvière Colliery	1194	1312	,, ,,	76 ,,	...	48
	,, ,,	1456	,,	,, ,,	81 ,,	...	49
	,, ,,	(1194—1456)	...	(5)	(5)	...	(52)
	,, ,, a new shaft	1361	...	3¼ R	81 R	...	45
216	,, Cuesmes Colliery	1548	1531	,, ,,	79·7 ,,	...	53
	,, ,,	1680	706	,, ,,	82 ,,	...	54
	,, ,,	(1548—1680)	...	... ...	(2·2)	...	(58)
	FRANCE.						
145	**Anzin**, Valenciennes †	126	...	2 R	56·5 R	...	21
	,, Shaft No. 1	658	...	,, ,,	67·7 ,,	54	39
		(126—658)	...	... ...	...	(11·2)	(47)
24	**Littry**, Calvados‡	325	107	2 C	61 C	70	36
25	**Decise**, Nièvre§	561	...	,, ,,	72 ,,	73	33
23	**Carmeaux**, Tarn ‖	630	60	,, ,,	67 ,,	74·3	53
	NORTH AMERICA.						
59	**Eastern Virginia**, Mills's Pit	420	...	... W	63¶ W	...	74
60	,, Wills's Pit	570	...	... ,,	65·5 ,,	...	65
61	,, Midlothian Pit	780	...	... ,,	68·7 ,,	...	65
	,, (another account)	609	...	... ?	?	...	45
	,, ,,	780	...	... ?	?	...	55

* The dip here is also considerable, and the Coal-measures are overlaid by a thick mass (300 to 400 feet) of water-bearing Lower Cretaceous strata.
   † Observations were made during sinking, in holes inserted horizontally in side of shaft.
   ‡ Strata nearly horizontal—pit dry.
   § The strata here dip 2° S.W., and overlie crystalline rocks.
   ‖ A new and dry pit; slow ventilation. The water of a well immediately above this pit, 38 feet deep, had a temperature of 55·5° F.
   ¶ Temperature of water collected at bottom of pit. Prof. Lebour's later statement relating to Virginian coal-pits seems more reliable.

TABLE III.—MINES OTHER THAN COAL. (For other particulars see Table I.)

Those observations which were made in Springs issuing from the rock or lode are marked S in Column V.; those in water collected in Wells or Sumps, W, and those in holes drilled in the Rock, R.

The Thermometric Gradient in Column VII. refers to the depths given in Column IV.; but where the figures are in *italics* and between brackets they refer to the gradients for the intermediate depths (also between brackets in Column IV). The difference of temperature between intermediate depths is given (in brackets) in Column VI.

I	II	I	IV	V	VI	VII
Number in Table I.	Name of Mine and Place.	Nature of Rock.	Depth of Mine.	Depth and Position of Hole. Rock R. Spring S. Water W.	Temperature at Depth.	Thermometric Gradient.
			Feet.	Feet.	Fahr.	Feet.
	ENGLAND.					
27	**Cornwall**, Dolcoath Mine *........	*Granite*	1440	... S	82° †	45
73	,, ,, (1822)...	,, ...	1380	? R	76	53
75	,, ,, (1853)...	,, ...	1632	,, ,,	79·5 ‡	55
74	,, ,, (1857)...	,, ...	1632	,, ,,	73	71
220	,, ,, ...	*Killas (slate)*	252	... ,,	64	18
	,, ,, ...	*Granite*	876	3 ? ,,	67·8	49
	,, ,, ...	,, ...	2124	... ,,	83	64
	,, ,, ...	,, ...	2244	... ,,	90	56
	,, ,, ...	....	(876—2124)	... ...	(*15·2*)	(*59*)
	,, ,, ...	....	(876—2244)	... ...	(*22·2*)	(*53*)
19	,, ,, ...	*Slate*............	240—300	3 ,,	58	34
	,, ,, ...	*Granite*	1320—1380 §	3 ,,	78	48
	,, ,, ...	....	1380—1440	... W	82	44
122a	,, Tresavean............	....	2130	... R	99	43·5
20	,, Huel Vor	*Slate*............	480—540	... W	63	42
	,, ,, ...	,, ...	780—840	... ,,	69 ǁ	45
21	,, Huel Damsel	*Granite*	300—360	... air	.. ¶	33
	,, ,, ...	,, ...	840—900	... ,,	... **	46
72	,, Botallock §............	*Granitic and horn-blendic rocks*	1128	1½ R	79	40
49f	,, Wheal Trenwith	*Slate*............	180	... S	55·5	40
	,, ,, ...	,, ...	660	... ,,	66	44
49g	,, South Roskear, Cambourne ††	,, ...	702	... ,,	62	59
	,, ,, ,, ...	,, ...	834	... ,,	71	40
49h	,, North Roskear	,, ...	822	... ,,	73	36
49i	,, East Pool............	,, ...	373	... ,,	59	41
49j	,, Wheal Uny, Redruth‡‡	,, ...	486	... ,,	61·5	48

* The Slate in this mine extends to depths of about 800 to 900 feet, the deeper levels are in Granite. This, the deepest of the Cornish mines, has now (1884) reached a depth of 2,400 feet.
† Air in gallery 80°.   ‡ Air in gallery 78°.
§ In the list given in page 218, the mean between these two depths (1,350 ft.) is taken.
ǁ Air in gallery 72°.   ¶ Air in gallery 61°.   ** Air in gallery 76°.
†† One level extends about 2000 feet under the sea.
‡‡ Thermal springs have been met with in several of the mines in this district.

I	II	III	IV	V	VI	VII
Number in Table 1.	Name of Mine and Place.	Nature of Rock.	Depth of Mine.	Depth and Position of Hole. Rock R. Spring S. Water W.	Temperature at Depth.	Thermometric Gradient.
	ENGLAND—*continued.*		Feet.	Feet.	Fahr.	Feet.
49k	**Cornwall**, Chacewater	*Slate*	768	... S	72	35
49l	,, East Wheal Virgin, Consolidated Mines...	,,	1509	... ,,	86·5	40
	,, ,, ,, ...	,,	1722	... ,,	92	39
49m	,, Wheal Towan, St. Agnes	,,	924	... ,,	72	42
49n	,, Wheal Prudence, St. Agnes	,,	654	... ,,	65·5	42
49a	,, East Wheal Crofty	,,	480	... ,,	61	44
	,, ,,	,,	810	... ,,	70·7	39
49b	,, Consolidated	,,	1704	... ,,	89	44·5
	,, ,,	,,	1764	... ,,	92·5	41·5
49c	,, United Mines	,,	1080	... ,,	74	45
	,, ,,	,,	1260	... ,,	89·5	32
49d	,, Great Wheal Fortune.	,,	840	... ,,	70	44
	,, ,,	,,	864	... ,,	73	39
49e	,, Marazion	,,	222	... ,,	56·3	43
	,, ,,	,,	480	... ,,	63	40
	,, ,,	,,	600	... ,,	66	40
70	,, Par Consols	,,	768	... R	74	34
71	,, ,,	,,	1248	... ,,	84	38
51	**Devonshire**, Wheal Friendship.	,,	282	... S	55	56
	,, ,,	,,	810	... ,,	69·5	41·5
	,, ,,	,,	(282—810)	... ...	(14·5)	(50)
75	**Cornwall**, Huel Alfred	,,	930	... S	70	47
28	,, ,, Trumpet	*Granite.*	768	... ,,	65	51
38	,, Levant	*Slate and Granite.*	1380	... R	89	46
76	,, ,,	,,	1530 *	... ,,	74	67
77	,, ,,	,,	1530	... ,,	85	45
76a	,, ,,	,,	1530	... ,,	87	42·5
40	,, Consolidated	*Granite*	1740	... ,,	85·3	49
39	,, Tresavean	,,	1572	... ,,	82	49
78	,, ,,	,,	1572	... ,,	82·5	48·5
79	,, ,,	,,	2112	... ,,	90·5	52
	,, ,,	,,	2112	... S	93·5	48
50	,, Binner Downs	*Slate*	300	... ,,	56·5	46
	,, ,,	,,	756	... ,,	67	44
	,, ,,	,,	1056	... ,,	82	33
			(300—1056)	... ...	(25·5)	(50)
80	,, United Consols	*Slate*	1728	... ...	93	41
82	,, ,,		1530	... ...	116	28
52	,, St. Ives Consols †	*Granite*	108	... S	57	18
	,, ,,	,,	462	... ,,	60·5	49
	,, ,,	,,	810	... ,,	71	41
	,, ,,	,,	(162—810)	... ...	(10·5)	(33)

* The station here was near the bottom of the shaft.
† The levels of many of the mines in this district extend beneath the sea.

ON UNDERGROUND TEMPERATURES—TABLE III

I	II	III	IV	V	VI	VII
Number in Table I.	Name of Mine and Place.	Nature of Rock.	Depth of Mine.	Depth and Position of Hole. Rock R. Spring S. Water W.	Temperature at Depth.	Thermometric Gradient.
	ENGLAND—*continued*.		Feet.	Feet.	Fahr.	Feet.
53	Cornwall, Wheal Wreath, St. Ives	Granite	162	... 8	53°	81
	,, ,, ,, ...	,, ...	1242	... ,,	70	65 *
	,, ,, ,, ...	,, ...	1482	... ,,	76	59 †
	,, ,, ,, ...	,, ...	(162—1242)	... ...	(17)	(64)
	,, ,, ,, ...	....	(1242—1482)	... ...	(6)	(40)
	,, ,, ,, ...	....	(162—1482)	... ...	(23)	(57)
18	,, Average of six mines.	Granite and Slate.	500—550	... W	65	35
	,, ,, ,, ...	,, ...	700—750	... ,,	65	43
	,, ,, ,, ...	,, ...	900—950	... ,,	71	44
	,, ,, ,, ...	,, ...	1350—1400	... ,,	79	47·5
54	,, and Devon,‡ Various mines in ten districts.	,, ...	180	... ...	54·8	37
		,, ...	432	... ...	66·8	40
	,, ,, ,, ...	,, ...	762	... ...	67·4	44
	,, ,, ,, ...	,, ...	1038	... ...	78·6	36
	,, ,, ,, ...	,, ...	1440	... ...	85·5	40
	,, average of 53 mines.	,, ...	354	... ...	...	35·4
	,, ,, ...	...	438	... ...	...	43·8
	,, ,, ...	...	684	... ...	...	64·2
130	Allanheads, Northumberland...	Carbs. limestone..	440	... W	57·3	35
15	Talargoch, Flintshire............	,,	660 §	2 R	54	132
	,, ,, ...	,,	1041 ‖	,, ,,	60·8	88
	,, ,, ...	,,	636 ¶	... ,,	58·8	64
	,, ,, ...	....	(660—1041)	... ...	(6·8)	(56)
160	,, ,, ...	,,	600 **	2 R	62	51
172	Sark, Island of............	Syenite............	324	... W	55·5	81
			384	... 8	57·2	59
				... ,,	58	
175	Ireland, Wicklow............	Silurian schists ...	552	... W	55·5	92
176	,, Waterford	Palæozoic schists.	672	... ,,	57·4 ††	90
177	,, County Cork............	Carboniferous shales	840	... ,,	61·5	80
	FRANCE.					
1	Giromagny, Belfort, Vosges	Porphyritic rocks.	332	... ...	53·6	50
	,, ,, ,, ...	,,	1420	... ...	73	50
	,, ,, ,, ...	,,	(332—1420)	... ...	(19·5)	(56)
12	Huelgoet, Brittany............	Silurian slates‡‡..	239	... W	54	75
	,, ,, ...	,,	781	... ,,	66	49
	,, ,, ...	,,	(239—781)	... ,,	(12)	(40)
11	Poullaouen ,, ...	,,	459	... 8	58·8	66
	SWITZERLAND.					
2	Bex, near Lausanne	Metamorphic rocks	721	... ...	65·5	30

* Temperature of lode. † Temperature of rock.
‡ The mines in this district are many of them very old.
§ Station 120 yards distant from shaft.
‖ Ditto 190 ditto
¶ Ditto 840 ditto
** Ditto 400 ditto
†† Air in gallery 57°.
‡‡ The slates are associated with quartzites and greenstone.

I	II	III	IV	V	VI	VII
Number in Table I.	Nature of Mine and Place.	Nature of Rock.	Depth of Mine.	Depth and Position of Hole. Rock R. Spring S. Water W.	Temperature at Depth.	Thermometric Gradient.
	AUSTRIA AND GERMANY.		Feet.	Feet.	Fahr.	Feet.
8	**Freiberg**, Himmelfahrt Mine	Gneiss*	870	... W	58°†	72
10	,, Junghöhe-Birke ,,	,,	656	... ,,	57·2	50
	,, ,, ,,	,,	1082	... ,,	62·5	58
	,, ,, ,,	,,	936	... ,,	61	55
13	,, Alte Hoffnung Gottes ,,	,,	200	... R	48·2	40
	,, ,, ,,	,,	886	... ,,	59	55
	,, ,, ,,	,,	1246	... ,,	66	54
	,,	,,	(886—1246)	... ...	(7)	(51)
7	,, Beschertglück Mine	,,	722	... R	54·5	63
	,,	,,	984	... ,,	60	58
			(722—984)	... ...	(5·5)	(48)
139	**Przibram**,‡ Bohemia	Silurian schists	621	2 R	56·7	
	,,		1202	,, ,,	58·3	
	,,		1900	,, ,,	61·4	126
	,,		(621—1202)	... ...	(7·6)	(72)
	,,		(621—1900)	... ...	(10·7)	(120)
149	**Schemnitz**, Hungary	Tertiary strata and syenite §	1047	1½ to 2½ R	...	74
	AMERICA.					
3	**Guanaxuato**, Mexico	Clay slate	1640	... ...	98	45
4	**Cabrera** ,,	?	164	... ...	63	55
83	**Chili**, Colorado mine ‖	Jurassic limestone	288	2 R	64·8¶	
	,,	,,	909	,, ,,	67**	150
	,,	,,	1362	,, ,,	(71·5)††	90
	,,	,,	(909—1362)	... ...	74·5	(40)
170	,, Chanarcillo	,,	930	... W or 8	69·2	110
171	**Brazil**, Minas Giräes		318	... ,,	67·9	105
	,, Morro Velho	Clay slate	892	... ...	81	
	AUSTRALIA.					
165a	**Ballarat**, Victoria	Palæozoic rocks	760	3 R	72·5	

* The gneiss of this district alternates with mica-schists, and is traversed in many places by masses and dykes of granite.
† Air in gallery 59°.
‡ The workings communicate with a great drainage tunnel 1,260 feet above the sea-level.
§ These rocks are traversed by veins of rhyolite.
‖ Total depth of mine 1500 feet.
¶ Air in gallery 66°.
** Air in gallery 65°.
†† Air in gallery 76·5°.

TABLE IV.—ARTESIAN WELLS AND BORINGS. (For other particulars see Table I.)

Column V.—The sign + indicates that the water overflowed at the surface (the height of overflow is given where known); − indicates the water stood below the surface.
Column VI.—These numbers give the difference between the temperature of the surface and that of the depth, for which see Table I.: those in brackets are differences between the special depths given in Column IV., for which the rate of increase is given in brackets and *italics* in Column VII.

I	II	III	IV	V	VI	VII
Number in Table I.	Place.	Nature of the Strata.	Depth.	Rise of Water relatively to Surface.	Difference between temperature at surface and depth.	Thermometric Gradient.
			Feet.	Feet.	Fahr.	Feet.
	ENGLAND.					
35	Sheerness *	London Clay, Woolwich Sands and Thanet Sands	361	—	10·8	34
	,,	,,	450	—	12·8	35
129	Kentish Town, London.	London Clay and Sands ......325 ft. Chalk......................646 ,,	1100	−210	19·7	52·3
	,,	Upper Greensand and Gault. 143 ,,	(0—550)	...	(10·6)	(52)
	,,	Red Sandstones (Devonian?) 188 ,,	(550—1100)	...	(9·0)	(60)
230	Richmond, Surrey	Tertiary strata, 242 ft., Cretaceous, 898 ft., Oolite, 36 ft.	1176	...	20·4	57
231	,, ,,	Red Sandstone (Devonian?) 257 ft.	1337	+	25·9	51·5
142	Chiswick	London Clay, Sands, and Chalk	396	−60	8	49
165	Southampton	Tertiary strata, 400 ft., and Chalk, 810 ft.	1210	—	19·7	65
210	Newport, Isle of Wight.	Tertiary Clays and Sands	467	+6	11·0	43
143	Swinderby, Lincoln	Lias and Rhætic .... 140 ft., New Red Sandstone.........1259 ,,	2600	—	30	66
	,, ,,	Permian Marls, &c......... 372 ,,	(0—1000)	—	(20)	(50)
	,, ,,	Carboniferous Strata ...... 129 ,,	(1000—2000)	—	(10)	(100)
153	Bootle, Liverpool §	Trias (Red Sandstones and Marls)..	1302	—	8·5	153
	SCOTLAND.					
45	Carse of Falkirk ‡	Alluvial Beds and Coal-measures	270	+	5	54
45a	Midlothian, average of 11	,, ,, ,,	213—350	+	...	48
123	Blythswood, Glasgow	Boulder Clay and Coal-measures	347	—	6·69	52
	,, ,,	,, ,, ,,	(60—347)	...	(5·3)	(54)
124	South Balgray ,,	,, and Greenstone	525	—	12·52	42
	,, ,,	,, ,,	(120—525)	...	(10·0)	(40)
	FRANCE.					
209	Dunkirk	London Clay, Sands, and Chalk	426	+	1·8	235

* See Whitaker's *Mem. Geol. Survey*, vol. iv. Part I. pp. 423—571.
‡ Thirty-two years elapsed, and the well was abandoned, before the temperature observations were made. It was then partly blocked.
§ Springs and convection currents interfere with results of this well.
‖ Kennet House Pit.

I	II	III	IV	V	VI	VII
Number in Table I.	Place.	Nature of the Strata.	Depth.	Rise of Water relatively to Surface.	Difference between temperature at surface and depth.	Thermometric Gradient.
	FRANCE—continued.		Feet.	Feet.	Fahr.	Feet.
208	Bourbourg, Dunkirk	London Clay, Sands, and Chalk *	544	+	9°	62
213	Croix (Dept. au Nord)	Chalk and Carboniferous Limestone †	271	+	7	39
32	Lille (St. Venant)	Tertiary Sands and Chalk resting on Carboniferous Limestone †	329	+	6·7	49
33	Aire, near St. Omer ‡	Tertiary Clays and Sands	205	+	5·0	35
36	Paris, École Militaire	Lower Tertiary Strata and Chalk	567	−	10	54
37	Paris, Grenelle	Lower Tertiary Strata, 148 feet	1797	+	30·0	58
	,, ,,	Chalk ....1394 ,,	(1312)	...	(23·7)	(55)
	,, ,,	Clays, Greensand ...... 255 ,,	(1707)	...	(9·2)	(56)
221	Paris, Passy	Same Strata as at Grenelle	1924	+	31·4	60
48a	Paris, St. Ouen	Tertiary (middle) Strata	216	+	4·3	50
40	Alfort (Marne) ‡	Lower Tertiary Strata	177	+ ?	6	30
62	Meaux (Marne) ‡	,, ,,	230	+ ?	6	38
48	Troyes (Aube)	Chalk and Gault	410 §	−	7·8	53
42	Rouen, St. Sever	Chalk and Jurassic Strata ?	600	−	13	46
43	St. André (Eure)	Chalk and Greensand	830	−	17	59
31	Tours §	,, ,,	460	+	10·4	44
68	Creuzot, Torcy Colliery‖	Trias, 1312 ft. ; Coal-meas., 505 ft.	1817	−	32	57
69	Creuzot, Mouillonge ,,	Trias, 1217 ft. ; Coal-meas., 1460 ft.	2677	−	51	52
101	Rochefort	Triassic Beds	2812	+	56·5	50
102	Virac (Tarn) ‡	,,	971	+ ?	28·25	31
183	Arcachon (Gironde) ‡	Upper Tertiary Strata	413	+ ?	5·4	71
	BELGIUM AND HOLLAND.					
180	Brussels	Lower Tertiary Strata	215	− 5	3·7	51
181	Aerschot	,, ,,	453	+ 16	6·9	65
65	Mondorff, Luxembourg.	Lias, 177 ft. ; Red Marls, 675 ft.	1647 ¶	+	31	53
	,, ,,	Muschelkalk, 465 ft. ; Red Sandste.				
46	Cosseigne-les-Luxembourg **	1020 ft. ; Quartzose schists, 52 ft.	2362	...	34·4	69
		?	1105	+ ?	...	45
207	Ostend	London Clay and Sands, 656 ft. Chalk, 210 ft.	981	+ ††	21·60	45·4
	,,	Red Chalk and Sand........ 92 ft.	(0—567)	...	(9·0)	(63)
	,,	Silurian schists, 30 ft. ?	(567—981)	...	(12·6)	(39)

* The water came from the Landenian or Thanet Sands (at 544 feet), but the bore was continued to a further depth of 275 feet in the chalk.
† The water rises from the Carboniferous Limestone.
‡ No information is given how the observations at these pits were made.
§ The lower part of the bore-hole, which was carried to a total depth of 517 feet, was obstructed. M. Walferdin supposes the heat caused by the boring instruments may have had scarcely time to be dissipated, but he assumed a higher surface temperature.
‖ At Torcy the works had been suspended six months. At Mouillonge three days.
¶ An overflowing spring rose here.
** M. Biver gives the rate of increase, but not the surface temperature, nor particulars of how the observations were made.
†† Although the water overflowed, it was in such small quantity that it was found necessary to take the temperatures at depths. There were small springs at 567 and 981 feet.

I	II	III	IV	V	VI	VII
Number in Table I.	Place.	Nature of the Strata.	Depth.	Rise of Water relatively to Surface.	Difference between temperature at surface and depth.	Thermometric Gradient.
			Feet.	Feet.	Fahr.	Feet.
	SWITZERLAND.					
29	Pregny, near Geneva	Molasse	713	−20	14·7°	48·5
	,, ,, ...	,,	(0—200)	...	(5·2)	(40)
	,, ,, ...	,,	(200—650)	...	(9·5)	(68)
	ITALY.					
67	Conselica, Ferrara	Alluvial Beds	164	+6½	5·1	32
94	Reggio *	Pliocene Marls	2297	...	...	56
178	Venice, Villa Francesco	Alluvial Beds	118	+	2·4	57
179	Venice, Gasworks	,,	236	+	6·9	34
89	Naples, Largo Vittoria	Volcanic Tuffs and Tertiary Strata	909	+	11·7	78
90	,, Royal Palace	,, ,,	1460	+	8·1	181
	GERMANY AND AUSTRIA.					
34	Neu Salzwerk	Liassic and Triassic Strata	2038	+	39·7	50·5
64	Vienna	Tertiary Strata	617	+	10·8	57
168	Buda-Pesth	Tertiary and Triassic Strata †	2066	...	120·5	24
	,,	,,	(0—328)	...	(20·4)	(12)
	,,	,, ,,	(328—1640)	...	(51·6)	(23)
	,,	,, ,,	(1640—2066)	...	(53·0)	(26)
167	Artern, Thuringia *		1012	...	...	72
	Rudersdorff, Prussia	Triassic Strata	880	+	26	34
87	Rehme, Westphalia		2280	+	39·4	58
144	Sperenberg, Berlin	Gypsum and Anhydrite, 2·3 ft. Rhen. Rock Salt ...3769 ft. Rhen.	4052 ‡	−	67·2	52
	,, ,, ,,		(0—700)	...	(13·0)	(55)
	,, ,, ,,	,,	(700—1500)	...	(13·2)	(60)
	,, ,, ,,	,,	(1500—2100)	...	(12·1)	(51)
	,, ,, ,,	,,	(2100—3390)	...	(27·7)	(50)
182	Minden, Westphalia		2230	+	42·3	52
166	Pitzbuhl, Magdeburg *		495	...	...	48
	RUSSIA.					
144a	St. Petersburg §	Silurian Strata resting on Granite.	656	+	14·8	44
	INDIA.					
184	Pondicherry	Alluvial Beds	261	...	11·1	52
151	Manegaon	Coal-measures	310	−	9·5 ‖	33
			(60—310)	...	(3·7)	(68)

* No particulars of these wells are given beyond depth and rate of increase of temperature.
† This is really a hot spring, due to the effects of old (Miocene ?) volcanic action.
‡ The total depth extended to 4,052 Rhenish feet = 4,172 English feet. The temp. obs. stopped at 3,390 feet.
§ An earlier account (*Revue Géol.* for 1865) gives a depth of 525 feet and a temperature of 50·5°.
‖ This result is assuming the mean temperature to be 75·2°, which is that of Jabalpur, a neighbouring town nearly on the same level. In the *Brit. Assoc. Rept.* the rate of increase (68 ft.) is calculated from the temperature at the depth of 60 feet (81°).

I	II	III	IV	V	VI	VII
Number in Table I.	Place.	Nature of the Strata.	Depth.	Rise of Water relatively to Surface.	Difference between temperature at surface and depth.	Thermometric Gradient.
	AFRICA.*		Feet.	Feet.	Fahr.	Feet.
92	Algeria, Baraki	Marls and Gravels	426	+	10?	42
97	,, Messis	Tertiary Strata?	277	+	4'6?	51
95	Ghadame, Tripoli	?	394	+	11	36
88a	Sahara Desert (mean of several wells)	Tertiary Marls and Gravels	...	+	...	36
	NORTH AMERICA.‡					
66	Charleston, S. Carolina.	Eocene Strata, 708 feet, overlying 398 feet of Cretaceous Strata	1106 (100—400) (400—1106)	... ...	22 (4) (16)	50 (75) (44)
84	Columbus, Ohio	Upper Devonian Sandsts. and Shales	2575 (90—2575)	... ...	35 (35)	73 (71)
88	Louisville, Kentucky	Silurian Limestones and Sandsts	2086	+170?	27'8	75
127	St. Louis	Lower Carboniferous Shales (429 ft.) Sandst. (360 ft.), Limestones (2725 ft.)	3029† (3843)	— ...	34 (52)	56 (74)
169	Chicago ‡	Silurian Limestone and Sandst	751	+	9'1	88
212	Bothwell, Ontario	Corniferous Limestone (Devonian)	475	+	8?	60
172	Sheboyan, Wisconsin	St. Peter's Sandstone (Silurian)	1475	+104		
223	La Fayette, Indiana		213	+		
	SOUTH AMERICA.					
224	Buenos Ayres	Pampean Beds	255	+	7'1	36

* The surface temperature at the African wells is very uncertain.
† The observations at 3,029 feet were taken with great care, and were considered reliable. The discrepancy at the greater depth is at present unaccountable, unless it were due to convection currents.
‡ In this well an infiltration of water from a higher level was suspected.

SUPPLEMENTARY OBSERVATIONS.

Several of these are remarkable for their great depth and high temperature.

Locality	Nature of Work.	Height of Ground.	Depth.	Temperature at Depth.	Thermometric Gradient.	—
		Feet.	Feet.	Fahr.		
232 **Richmond,** Surrey*	Artesian well	17	{1337, 1447}	{75½°, 76¾}	{52·4, 54·1}	Brit. Assoc. Rep., 1885.
233 **Aberdare,** Cwmpennar.	Coal	800	1272	66½	60	,, ,, ,, ,,
234 **Schladebach,** Dürrenberg†	Artesian well	...	5630	...	65	,, ,, ,, 1889.
235 **Pittsburgh,** Dilworth well ‡	Slate rocks	900	4295	114	69·3 ?	,, ,, ,, ,,
236 **Wheeling,** Virginia §.	Oil well	...	4460	110·2	74·1	,, ,, ,, 1892.

* These observations were made subsequently to those of Professor Judd (No. 230). Diameter of bore-hole at top, 16¼ inches. Temperature of overflowing water, 59°.

† Diameter of bore-hole at top, 126 mm.; at bottom, it had diminished to the size of a man's little finger.

‡ Diameter at top was 6 inches. Gradient not considered reliable.

§ The well did not succeed and was abandoned; the observations were made subsequently.

INDEX TO NAMES OF PLACES IN TABLE I.

—	No. in Table I.	—	No. in Table I.
Aberdare, Upper Duffryn Colliery	112	**Cornwall,** Huel Trumpet	28
,, New Tredegar Colliery	113	,, Levant	38, 76, 77,
,, Dowlais Colliery	114		39, 78, 79,
,, Cwmbach	115	,, **Tresavean**	122a
Aerschot, Belgium	181	,, Consolidated	40, 49b
Aire, Pas de Calais	33	,, East Wheal Crofty	49a
Alfort, Marne	49	,, United Mines	49c
Allanheads Breckonhill	133	,, Great Wheal Fortune	49d
,, Northumberland	130, 131	,, Marazion	49e
,, Weardale	132	,, Wheal Trenwith	49f
Annesley Colliery	106	,, South Roskear, Camborne	49g
Anzin, Nord, France	145, 146, 147, 148	,, North Roskear	49h
		,, East Pool	49i
Arcachon, Gironde	183	,, Redruth, Wheal Uny	49j
Artern, Thuringia	167	,, ,, Chacewater	49k
Ashton Moss Colliery, Manchester	218	,, ,, Cons. Mines	49l
		,, St. Agnes	49m
Ballarat, Australia	165a	,, Wheal Prudence	49n
Baraki, Algiers	92	,, Binner Downs	50
Ben Tallah	91	,, St. Ives Consols	52
Bex, Switzerland	2	,, Wheal Wreath	53
Birmingham Waterworks	188	,, Par Consols	70, 71
Blackburn	187	,, Botallock	72
Blythswood, near Glasgow	123	,, United or Fowey Consols	80, 81, 82
Boldon, Newcastle	150	Cosseigne-les-Luxembourg	46
Bootle, Liverpool	153	Crawriggs, near Glasgow	134
Bothwell, Ontario, U.S.	160, 212	Creuzot (Torcy), Saône-et-Loire	68
Bourbourg, near Dunkirk	208	,, (Mouillonge)	69
Bradford, Yorkshire	186	Croft, Whitehaven	226, 227
Braintree	202	Croix, Dept. du Nord	213
Brazil, Minas Giraës	171		
Bristol, Kingswood	155, 156		
Brittany, Poullaouen	11	Deal Waterworks	196
,, Huelgoet	12	Decise, Nièvre	25
Brussels	180	Denton, Manchester	229
Buda-Pesth, Hungary	168	Deptford Waterworks	200
Buenos Ayres	224	Devonshire	51
		Dover Castle	197
Cabrera	4	Dover Waterworks	198
Carisbrooke Castle, Isle of Wight	23	Dukinfield	85, 86, 159
Carmeaux (Tarn)	204	Dundee	185
Carrickfergus, Belfast	125	Dunkirk	209
Charleston, U.S.A.	66	Durham, **Hetton Colliery**	116
Ciega	90	,, **S. Hetton Colliery**	196
Cheshire, Bredbury }	162	,, **Heaton Colliery**	117
,, Nook Pit }			
Chicago	169	**Eastbourne**	195
Chili	83	**Eastern Virginia,** U.S.A., Mills's Pit	59
Chili, Chanarcillo	170	,, ,, Wills's Pit	60
Chiswick, Middlesex	142	,, ,, Midlothian Pit	61
Colchester	205		
Columbus, Ohio, U.S.A.	84	**Freiberg, Saxony**	7, 8, 9, 10, 13
Conselica, Ferrara, Italy	67		
Cornwall	18, 47, 174	Ghadames, Tripoli	95
,, and Devon	94	Gironagny, near Belfort	1
,, Dolcoath	19, 27, 73, 74, 75, 220	Gosport	211
		Grimsby Docks	199
,, Huel Vor	20, 49p	Guanaxuato, Mexico	3
,, Huel Damsel	21		
,, Huel Alfred	26	Hucknall Torkard Colliery, Notts	105

	No. in Table I.		No. in Table I.
Ireland, various mines	173	Radstock, Somerset, May Pit	163
,, county Cork	177	,, ,, Ludlow Pit	164
,, ,, Waterford	176	Ram's Mine, Lancashire	132
,, ,, Wicklow	175	Reggio, Italy	94
		Rehme, Westphalia	87
Kentish Town	129	Richmond, Surrey	230, 231
Kidderminster	189	Rochefort, Charente	101
Killingworth	17	Rosebridge Colliery, Wigan	118, 12
Kiveton Park Colliery	107	Ruabon, N. Wales	110
		Rudersdorf, Berlin	41
La Fayette, Indiana	223		
Lille (St. Venant)	32	Sahara Desert	88
Littry, Calvados	24	St. Andre, Eure	43
Louisville, Kentucky	88	St. Gothard Tunnel	154
Lye Cross, Dudley	228	St. Helens Waterworks	190
		St. Louis, U.S. America	127
Manchester, Ashton Moss	161	St. Petersburg	96, 144
Manegaon, India	151	St Sever, Rouen	42
Meaux, Marne	62	Sark and Herm	172
Meiahsdalon, Algeria	98	Scarborough	194
Messis, Algeria	97	Schemnitz, Hungary	149
Miculpampa, Peru	6	Scotland, Carse of Falkirk	45
Minden, Prussia	182	Seraing, Liège, Marie Colliery	140
Moira Colliery, Warwickshire	109	Seraing, Liège, Henri Guillaume	141
Mondorff, Luxembourg	57, 65	Sharlston Colliery, Barnsley	119
Mons (Couchant de Flenu)	58	Sheboyan, Wisconsin	172
,, Cuesmes	216	Sheerness	35
,, La Louvrière	215	Sittingbourne	201
,, Grameries	214	Southampton Common	165
Mont Cenis Tunnel	55	South Balgray	124
Monte Massi, Tuscany	128	Sperenberg, Berlin	144
Montigny, Belgium	163	Stowmarket	138
Moscow	135	Sunderland	30
		Swanwick Colliery	108
Naples, Largo Vittoria	89	Swinderby, Lincoln	143
,, Royal Palace	90		
Neath, South Wales	22	Talargoch, Flintshire	158, 160
Neu Salzwerk	34	Tehuilotopec, Mexico	5
Neuffen, Wurtemberg	56	Tours	31
Newport, Isle of Wight	210	Tranmere, Cheshire	191
Norley Coal Co., Wigan	111	Trowbridge Waterworks	206
North Seaton, Newcastle	217	Troyes, Aube	48
Ostend, Belgium	63, 207	Venice Gasworks	179
Oued-el-Halleg	93	Venice, Villa Francisco Grande	178
		Victoria Colliery, Wakefield	120
Paris, École Militaire	36	Vienna	64
,, Grenelle	37	Virac (Tarn)	102
,, St. Ouen	48a		
,, Passy	221	Wallasey	192
,, La Chapelle St. Denis	187	Whitehaven	14
Pendleton Colliery, Manchester	104, 122	Wimbledon	203
Percy Main	16	Workington	15
Pitzbuhl, Magdeburg	166	Worksop	193
Pondicherry, E. India	184	Worthington Colliery, Lancashire	121
Pontypridd, S. Wales	152		
Pregny, near Geneva	29	Yakoutsk, Siberia	44
Przibram, Bohemia	139		

RICHARD CLAY AND SONS, LIMITED,
LONDON AND BUNGAY.

www.ingramcontent.com/pod-product-compliance
Lightning Source LLC
Chambersburg PA
CBHW030019240426
43672CB00007B/1013